Lecture Notes in Networks and Systems 1681

The series "Lecture Notes in Networks and Systems" publishes the latest developments in Networks and Systems—quickly, informally and with high quality. Original research reported in proceedings and post-proceedings represents the core of LNNS.

Volumes published in LNNS embrace all aspects and subfields of, as well as new challenges in, Networks and Systems.

The series contains proceedings and edited volumes in systems and networks, spanning the areas of Cyber-Physical Systems, Autonomous Systems, Sensor Networks, Control Systems, Energy Systems, Automotive Systems, Biological Systems, Vehicular Networking and Connected Vehicles, Aerospace Systems, Automation, Manufacturing, Smart Grids, Nonlinear Systems, Power Systems, Robotics, Social Systems, Economic Systems and other. Of particular value to both the contributors and the readership are the short publication timeframe and the world-wide distribution and exposure which enable both a wide and rapid dissemination of research output.

The series covers the theory, applications, and perspectives on the state of the art and future developments relevant to systems and networks, decision making, control, complex processes and related areas, as embedded in the fields of interdisciplinary and applied sciences, engineering, computer science, physics, economics, social, and life sciences, as well as the paradigms and methodologies behind them.

Indexed by SCOPUS, EI Compendex, INSPEC, WTI Frankfurt eG, zbMATH, SCImago.

All books published in the series are submitted for consideration in Web of Science.

For proposals from Asia please contact Aninda Bose (aninda.bose@springer.com).

Preface

This volume contains a selection of extended and peer-reviewed high-quality research papers presented at the XIX International Symposium of Organizational Sciences—SymOrg 2024, titled "Unlocking the Hidden Potential of Organization Through Merging of Humans and Digitals," held in Zlatibor.

The theme, "Unlocking the Hidden Potential of Organization Through Merging of Humans and Digitals," is inspired by recent advancements in digital technologies, particularly artificial intelligence. Over the past few years, digital technologies have been increasingly adopted across organizations in various sectors. The unique characteristics of modern digital technologies, when synergized with human capabilities, hold the capacity to unlock hidden organizational potential. However, achieving this synergy requires more than the mere adoption of advanced tools; it necessitates a multidimensional perspective, which is thoroughly explored in this volume.

The symposium brought together numerous scholars who presented 149 papers across 15 thematic areas. Following a rigorous peer-review process, this volume features 14 papers that represent significant contributions from the authors. The thematic areas covered include e-business, financial innovations, marketing and communication, human resource management, project management, and quality management and standardization. The aim of this volume is to share the latest research with both academia and practitioners in the field of organizational sciences.

This book was made possible thanks to the valuable contributions of domestic and international authors, the dedicated work of reviewers, the collaboration between academia and the broader community, and the support of all partners and individuals involved. We express our special gratitude to the Ministry of Science, Technological Development and Innovation of the Republic of Serbia. We are grateful to the publisher, Springer, for their efforts in making this volume accessible in both traditional and online formats.

December 2024

Velimir Štavljanin
Ivana Mijatović
Ivan Luković

Organization

Program Committee

Kostić-Stanković Milica (Chair)	University of Belgrade, Faculty of Organizational Sciences, Serbia
Abrudan Denisa	West University of Timisoara, Faculty of Economics and Business Administration, Romania
Begičević Ređep Nina	University of Zagreb, Faculty of Organization and Informatics, Croatia
Berber Nemanja	University of Novi Sad, Faculty of Economics in Subotica, Serbia
Bijakšić Sanja	University of Mostar, Faculty of Economics, Bosnia and Hercegovina
Bogdanović Zorica	University of Belgrade, Faculty of Organizational Sciences, Serbia
Bogojević Arsić Vesna	University of Belgrade, Faculty of Organizational Sciences, Serbia
Brkić Nenad	University of Sarajevo, Faculty of Economics, Bosnia and Hercegovina
Ćetković Jasmina	University of Montenegro, Faculty of Economics, Montenegro
Čudanov Mladen	University of Belgrade, Faculty of Organizational Sciences, Serbia
Damnjanović Vesna	University of Belgrade, Faculty of Organizational Sciences, Serbia
Delibašić Boris	University of Belgrade, Faculty of Organizational Sciences, Serbia
Despotović-Zrakić Marijana	University of Belgrade, Faculty of Organizational Sciences, Serbia
Dlačić Jasmina	University of Rijeka, Faculty of Economics, Croatia
Domanović Violeta	University of Kragujevac, Faculty of Economics, Serbia
Drezgić Saša	University of Rijeka, Faculty of Economics, Croatia
Đukić Suzana	University of Niš, Faculty of Economics, Serbia
Filipović Jovan	University of Belgrade, Faculty of Organizational Sciences, Serbia

Hertzler Eric	Université Paris-Est Créteil, IAE Paris-Est School of Management, France
Janković Marija	University of Paris-Saclay, CentraleSupélec, France
Jaško Ondrej	University of Belgrade, Faculty of Organizational Sciences, Serbia
Klačmer Čalopa Marina	University of Zagreb, Faculty of Organization and Informatics, Croatia
Knežević Miloš	University of Montenegro, Faculty of Civil Engineering, Montenegro
Kragulj Dragana	University of Belgrade, Faculty of Organizational Sciences, Serbia
Kuzmanović Marija	University of Belgrade, Faculty of Organizational Sciences, Serbia
Kuljanin Goran	DePaul University, Driehaus College of Business, USA
Lazarević Saša	University of Belgrade, Faculty of Organizational Sciences, Serbia
Lečić-Cvetković Danica	University of Belgrade, Faculty of Organizational Sciences, Serbia
Llorens Ariadna	Universitat Politècnica de Catalunya, Spain
Luković Ivan	University of Belgrade, Faculty of Organizational Sciences, Serbia
Ljepava Nikolina	American University in the Emirates, College of Business Administration, Dubai
Marič Miha	University of Maribor, Faculty of Organizational Sciences, Slovenia
Marton Peter	University of Žilina, Faculty of Management Science and Informatics, Slovakia
Mazzara Manuel	Innopolis University, Faculty of Computer Science and Engineering, Russia
Mihić Marko	University of Belgrade, Faculty of Organizational Sciences, Serbia
Miladinović Slobodan	University of Belgrade, Faculty of Organizational Sciences, Serbia
Mirakovski Dejan	University Goce Delcev, Faculty of Natural and Technical Science, North Macedonia
Petrović Dejan	University of Belgrade, Faculty of Organizational Sciences, Serbia
Petrović Nataša	University of Belgrade, Faculty of Organizational Sciences, Serbia
Pizzi Gabriele	Università di Bologna, Alma Mater Studiorum, Italy
Pokimica Jelena	Boise State University, USA

Reviewers

Brđanin Dražen	University of Banja Luka, Faculty of Electrical Engineering, Banja Luka, Bosnia and Herzegovina
Čudanov Mladen	University of Belgrade, Faculty of Organizational Sciences, Serbia
Čular Marko	University of Split, Faculty of Economics, Croatia
Danilović Miloš	University of Belgrade, Faculty of Organizational Sciences, Serbia
Dlačić Jasmina	University of Rijeka, Faculty of Economics, Croatia
Erdoğan Nevnihal	Faculty of Architecture and Design, Kocaeli University, Turkey
Filipović Jovan	University of Belgrade, Faculty of Organizational Sciences, Serbia
Fırat Hüsniye	Beykoz University, Faculty of Business and Management Sciences, Turkey
Ham Marija	J. J. Strossmayer University of Osijek, Faculty of Economics, Croatia
Jednak Sandra	University of Belgrade, Faculty of Organizational Sciences, Serbia
Kalinić Zoran	University of Kragujevac, Faculty of Economics, Serbia
Karanović Goran	University of Rijeka, Faculty of Tourism and Hospitality Management, Croatia
Kelić Ivan	J. J. Strossmayer University of Osijek, Faculty of Economics, Croatia
Kljajić Boštnar Mirjana	University of Maribor, Faculty of Organization and Management, Kranj, Slovenia
Knežević Snežana	University of Belgrade, Faculty of Organizational Sciences, Serbia
Kosar Tomaž	University of Maribor, Faculty of Electrical Engineering and Computer Science, Maribor, Slovenia
Kostić-Stanković Milica	University of Belgrade, Faculty of Organizational Sciences, Serbia
Kovačević Biljana	University of East Sarajevo, Faculty of Business Economics Bijeljina, Republic of Srpska, Bosnia and Herzegovina
Kovačević Ivana	University of Belgrade, Faculty of Organizational Sciences, Serbia
Krivokapić Jovan	University of Belgrade, Faculty of Organizational Sciences, Serbia
Kuljanin Goran	DePaul University, Driehaus College of Business, USA

Labus Aleksandra	University of Belgrade, Faculty of Organizational Sciences, Serbia
Marinković Sanja	University of Belgrade, Faculty of Organizational Sciences, Serbia
Meško Maja	University of Maribor, Faculty of Organizational Sciences, Slovenia
Miladinović Slobodan	University of Belgrade, Faculty of Organizational Sciences, Serbia
Milić Tanja	University of Belgrade, Faculty of Organizational Sciences, Serbia
Petrović Nataša	University of Belgrade, Faculty of Organizational Sciences, Serbia
Popović Aleksandar	University of Montenegro, Faculty of Natural Sciences and Mathematics, Montenegro
Poustourli Katerina	International Hellenic University, Greece
Radovanović Sandro	University of Belgrade, Faculty of Organizational Sciences, Serbia
Rožanc Igor	University of Ljubljana, Faculty of Computer and Information Science, Ljubljana, Slovenia
Samardžija Luka	University of Rijeka, Faculty of Economics and Business, Croatia
Simić Dejan	University of Belgrade, Faculty of Organizational Sciences, Serbia
Stošič Biljana	University of Belgrade, Faculty of Organizational Sciences, Serbia
Stubelj Igor	University of Primorska, Faculty of Management, Slovenia
Šegota Tina	University of Maribor, Faculty of Tourism, Maribor, Slovenia
Šošević Uroš	University of Belgrade, Faculty of Organizational Sciences, Serbia
Tomičić Furjan Martina	University of Zagreb, Faculty of Organization and Informatics, Varaždin, Croatia
Vasiljević Dragan	University of Belgrade, Faculty of Organizational Sciences, Serbia
Vukmirović Dragan	University of Belgrade, Faculty of Organizational Sciences, Serbia
Vukovič Goran	University of Maribor, Faculty of Organizational Sciences, Maribor, Slovenia
Žitnik Slavko	University of Ljubljana, Faculty of Computer Science and Informatics, Slovenia

Sponsors

Banka Intesa

Nelt

Netconomy

Ministry of Science, Technological Development and Innovation

Knjaz Miloš

GCF Group

Calvet

Zlatiborac

Omilab

Nescafe

Contents

About the Editors

Velimir Štavljanin is a Full Professor at the Faculty of Organizational Sciences, University of Belgrade. His research interests include marketing management, digital technologies in marketing, online consumer behavior, and multimedia technologies. He has authored or co-authored more than 20 books and chapters, and over 120 papers. He is a reviewer and editorial board member for various national and international journals. He is the editor-in-chief of the Management: Journal of Sustainable Business and Management Solutions in Emerging Economies. He is a member of the Serbian Marketing Association and the American Marketing Association. He is the chairman of the Supervisory Board of Serbian Marketing Association (SeMA). His teaching areas are marketing management, brand management, consumer behavior, digital marketing, and performance measurement in marketing and communications. He currently serves as the chair of the B.Sc. program in Marketing Management and Communications at the Faculty of Organizational Sciences. He has taken part in more than 20 commercial projects for organizations in Serbia and the surrounding region. Additionally, he has contributed to several research projects funded by the Government of the Republic of Serbia. As a part of these projects, he has been involved in the implementation of various technical solutions.

Ivana Mijatović is a Full Professor at the Faculty of Organizational Sciences, University of Belgrade. She is the leader of partner teams in the Horizon Europe projects HSbooster.eu (Standardisation Booster for H2020 & HE research results), Edu4Standards.EU (Education for Standardisation in the EU), and IAM4RE (Intellectual Asset Management For Research And Entrepreneurship). In the Horizon Europe project StandICT2026.eu, she is the chair of the EUOS group for education in ICT standardization. Since 2012, she has been the vice president of the board of the European Academy for Standardization (EURAS, https://www.euras.org/). In 2018/2019, she was the president of the International Cooperation in Education about Standardization Committee (ICES, http://www.standards-education.org/). In 2017/2019, she was a member of the Joint Initiative on Standardization (JIS Action 3) working group of the European Union. Since 2015, she has been a member of the STARTed Team (Expert Team on Standardization and Regulatory Techniques-Education in Standardization) of the United Nations Economic Commission for Europe (UNECE). She is a member of the national committee for standardization KS I1/07-Software Engineering, IT for Education and the Internet, at the Institute for Standardization of Serbia. She served as a section editor in the Encyclopedia of Sustainable Management (Springer Nature, 2023), co-editor of the book Sustainable Development: Knowledge and Education about Standardization (Springer Nature, 2020) and author of four chapters in the book series Springer Nature "CSR, Sustainability, Ethics & Governance". A list of her publications can be found at https://www.researchgate.net/profile/Ivana_Mijatović.

Ivan Luković received his diploma degree (5 years) in Informatics from the Faculty of Military and Technical Sciences in Zagreb in 1990. He completed his M.Sc. (former Mr, 2 years) degree at the University of Belgrade, School of Electrical Engineering in 1993, and his Ph.D. at the University of Novi Sad, Faculty of Technical Sciences in 1996. Currently, he works as a Full Professor at the Faculty of Organizational Sciences of the University of Belgrade, where he lectures in several computer science and informatics courses. His research interests are related to database systems, information systems, business intelligence, software engineering, and data science. He is the author or co-author of over 200 papers, 4 books, and 30 industry projects and software solutions in the area. He supervised 12 completed Ph.D. theses. He created a new set of B.Sc. and M.Sc. study programs in information engineering, i.e., Data Science, at the Faculty of Technical Sciences. The programs were accredited for the first time in 2015. Currently, he is a chair of Managing Board of the Computer Science and Information Systems (ComSIS) journal, and a chair of M.Sc. study program in Information Engineering at Faculty of Organizational Sciences. He is also a member of Serbian AI Society.

Competition Between Ecosystems in the Digital Economy: Challenges and Policy Adaptations

Siniša Milošević[1(✉)] and Željko Spasenić[2]

[1] Commission for Protection of Competition, Belgrade, Republic of Serbia
sinisa.milosevic3@gmail.com
[2] University of Belgrade, Faculty of Organizational Sciences, Belgrade, Republic of Serbia

Abstract. This paper explores how traditional competition theories need to evolve in the context of **digital ecosystems (DEs)**. By analyzing the structural dynamics and competition mechanisms in well-known digital platforms, including Apple's iOS, Google's Android, and Microsoft's integrated services, we identify the limitations of current competition policies and propose necessary adjustments for regulating DEs effectively. Our findings suggest that existing frameworks are insufficient for addressing the complexities of DEs, and we offer policy recommendations tailored to the **digital economy**'s unique challenges.

Keywords: digital ecosystem · digital economy · competition policy

1 Introduction

The digital economy has transformed traditional market dynamics, giving rise to new business models centered on digital ecosystems (DEs). DEs are complex networks where platforms, services, and users interact to create value. However, the growth and influence of these ecosystems challenge existing competition policies, which were designed with traditional markets in mind. This paper explores these challenges and proposes adaptations to competition laws that can better address the rapid evolution of the digital marketplace.

In the era of the digital economy, the concept of digital ecosystem (DE) has emerged as a framework for the analysis and understanding the complex interplay between various actors, technologies, and markets. As businesses tend to digitalize their operations (Kronblad & Envall Pregmark, 2021), the dynamics of competition have been significantly reshaped, giving rise to a new market model where traditional industry boundaries blur, and DEs become drivers of innovation and value creation.

The development and sustainability of DEs have been linked to the broader concept of the digital economy, which encompasses the dynamic interactions and competition among various entities in the digital space (Ilchenko, 2021). As companies increasingly transition their activities to digital platforms and ecosystems to maintain global competitiveness, the economic implications of these transformations become a critical area of study (Lipovenko et al., 2022).

V. Štavljanin et al. (Eds.): SymOrg 2024, LNNS 1681, pp. 1–16, 2025.
https://doi.org/10.1007/978-3-032-08090-5_1

Hein et al. (2019) define DE as a network of organizations, individuals, and technologies that interact and co-evolve to create value through digitally enabled processes. DEs appears in various forms, from e-commerce marketplaces to social media platforms, each characterized by its unique mixture of participants and usually complex business interactions. The infrastructure of DEs is purposively developed to facilitate the production, marketing, distribution and sale or delivery of various products and services.

Petropoulos (2020) distinguishes between two groups of DEs: (i) aggregators and (ii) marketplaces. Aggregators are platforms that offer end users valuable services, along with facilitating interactions with third-party producers or service providers. Marketplaces are platforms whose core business is to create efficient and purposeful connection between consumers and suppliers of goods and services. Typical representatives for the first group are online search engines (e.g. Google) and social media platforms (e.g. Facebook, Instagram) while the second group is represented by online marketplaces such as booking.com, Amazon and Google Play.

The phenomenon of DEs has gained attention in various economic theories, with the primary aim to understand the process of value creation, innovation, and competition within these complex networks (Suseno et al., 2018). Economic theories, such as **platform economics** (Spulber, 2018) and **two-sided markets** theory (Filistrucchi et al., 2014) offer alternative perspectives that better capture the dynamics of DEs. These theories emphasize the significance of value creation through interactions among versatile stakeholders and the role of digital innovation capabilities in driving economic growth (Suseno et al., 2018; Xie et al., 2023).

The existing fund of knowledge on DEs within the digital economy is multifaceted, encompassing various dimensions such as value creation, innovation, entrepreneurship, and competition. In this research paper, we aim to explore the **market competition** between DEs within the context of the digital economy.

2 Theoretical Background

Traditional competition theories, focused on monopoly and oligopoly behaviors, do not capture the multi-sided, network-driven competition in DEs. Digital platforms, such as iOS and Android, operate on the basis of indirect **network effects** where the value to users increases as more participants join the ecosystem. This requires an adjustment to market definitions and policy frameworks that were originally intended to regulate linear, product-based markets.

Market competition refers to the rivalry between firms operating within the same market, competing for market share, consumers, and profitability through various strategies such as pricing, product differentiation, and marketing efforts (Yu & Liu, 2022). Historically, at the beginning of **competition policy** development, monopolies were identified as the biggest threat for market competition (Bunn, 1949). Competition policy comprises a set of rules that regulate competition on the market. In that context, competition policy should protect the competition process, not individual market participants. The market competition within DEs attracted substantial attention in literature and practice. Current fund of academic knowledge provides vague conclusions regarding DEs impact on market competition and the need to significantly adjust **antitrust law** to address the challenges imposed by them.

To fully grasp the concept of competition, it is important to distinguish it from competitiveness. Competitiveness pertains to various levels—including individual companies, specific economic sectors, regions, or entire national economies—each with its own definition (Snowdon and Stonehouse, 2006). Competition in the market occurs when two or more companies strive to increase the sales of their products or services, whether through pricing strategies or adjustments in quantity. This rivalry creates competitive pressure, which enhances the economic efficiency of businesses within the market. Market competition can be seen as a process where companies vie for the attention of final consumers by offering higher-quality products and services at lower prices to become the preferred choice. This competitive interaction induces pressure that drives companies to improve their effectiveness and boost productivity. As productivity increases, production costs tend to decrease. Competitive pressure constantly encourages companies to be more productive than others. To be more productive than others means to be able to offer products at lower prices and achieve a larger market share, and thus to become more competitive compared to others.

Competition policy is generally divided into two main components: state aid and competition. The state aid aspect involves competition rules designed to prevent governments from interfering in the competitive market process. This research focuses more extensively on the competition component. This section is based on competition rules that are quite similar across different jurisdictions worldwide. These rules are typically outlined in each country's competition laws. The competition part of competition policy is founded on three fundamental pillars: restrictive agreements, abuse of dominance, and mergers (Milošević, 2023).

The emergence of digital ecosystems has significantly altered traditional market dynamics, leading to the adoption of new competitive strategies, and posing challenges for competition law frameworks (Lianos, 2022). Multi-sided platforms and ecosystems challenge conventional views of competition by creating value chains that extend across different market segments, thereby modifying competitive strategies and market configurations (Rochet & Tirole, 2003). Jacobides & Lianos (2021) stressed the important role of DEs in competition law in theory and practice and offered a comprehensive understanding of the legal and regulatory dimensions influencing market competition within DEs.

The main goal of competition policy is to secure fair and appropriate competition between current and potential market players (Lončar & Milošević, 2013). The other frequently stated objectives of competition policy are dispersion of economic power and economic integration (Niels et al. 2016). Unrestricted entry of new market participants and the absence of barriers to entry intensify competitive pressure.

The rapid advancement of communication technologies, information technology, and the Internet has revolutionized the exchange of goods and services in the market. This technological progression has led to the evolution of new business models adopted by various market participants, with the primary goal of facilitating interactions between individuals and companies. In the realm of digital ecosystems, these developments have been particularly impactful. Digital ecosystems harness these innovations to create interconnected networks that enhance connectivity, reduce transaction costs, and promote seamless collaboration among stakeholders. These new business models focus on leveraging platforms and technologies to foster more efficient and dynamic interactions, thereby driving growth and innovation within the digital marketplace (Perić, et al., 2022).

Hylton (2019) believes that the significant market dominance of large platforms is mostly attributed to economies of scale rather than anticompetitive behavior. According to the author, there is no distinct feature of digital platforms warranting a revision of antitrust regulations and should not impede the enforcement of antitrust legislation. The similar conclusions are shared by Veljanovski (2022) who thinks that there is no significant and urgent need for the reform of competition law.

Traditional competition theories often focus on market dominance, barriers to entry, and the behavior of firms with significant market power. However, in DEs, these theories need adjustments to address the digital market dynamics. Innovative technology development, rise of the internet, big data, artificial intelligence, machine learning, and block chain are strongly changing industry of DEs.

Technological advancements have significantly disrupted the industry of digital ecosystems. These ecosystems, with their innovative business models, have introduced more efficient and diversified products, services, and channels into the digital sector. The disruption has been profound and continues to reshape the industry by fostering a more competitive environment, marked by the emergence of new players such as startups and platforms. Digital ecosystems offer faster and more cost-effective solutions for customers, gaining substantial popularity, especially among those who were previously underserved by traditional companies. This trend is particularly pronounced among micro, small, and medium-sized enterprises, which have greatly benefited from the accessible and tailored offerings provided by digital ecosystems (Barjaktarović et al. 2022). The openness of digital infrastructures, the governance of platform interactions, and the enforcement of fair play are all areas where traditional theories may fall short.

According to Petropoulos (2020), competition issues in DEs often involve a star network structure, where each user is connected to a central hub or platform. This platform supplies the infrastructure and helps users interact with each other. To manage this, platforms do three main thing: they use open digital infrastructures, that let various stakeholders to manage their services and content; they create rules for governance and put resources into enforcing these rules to ensure a balance between controlling the platform and motivating users to participate actively and create value; they establish fair and effective systems for resolving disputes to address any unfair practices quickly.

Key features of DEs can lead to economic dynamics where platforms gain more control over the ecosystem and cause the market to tip in favor of one player. In multi-sided platforms, one side of the market can gain extra value from interacting with the other side (Kapoor, 2018). By attracting potential customers on the one side, platforms can get higher benefit from advertising. CMA (2020) illustrates the high concentration of advertising revenues in platforms such as Google and Facebook. Digital goods and services are typically produced at a significant fixed cost but no or insignificant variable cost (Varian et al., 2004). Once set up, digital companies can quickly grow by reaching more users with minor marginal cost of customer acquisition.

Machine learning and artificial intelligence have greatly increased the value of data for digital companies. By gathering, analyzing, and combining large amounts of data, companies can enhance their products and expand into new areas. Since machine learning works better with more data, companies with access to large data sets can improve their services more than those with limited data (Martens, 2020). The value a user gets from joining a platform can increase as more users join the same platform. This leads to a wider and better range of products and services, making the platform even more valuable and attracting more users and developers in a positive feedback loop.

The interaction of above mention forces can make the ecosystem more economically dependent on the platform that facilitates most interactions between different users. The platform that first attracts many users can dominate the ecosystem. In other words, in DEs, we often see not just competition within the market, but competition for the market (Iossa et al. 2022). Defining the relevant market is a fundamental concept that underlies all critical issues in competition policy. It serves as a widely used analytical framework and is often the initial step in analyzing competitive conditions and investigating potential competition infringements. Accurately determining the boundaries of the relevant market enables the identification of the market structure and its participants, as well as the assessment of the competitive pressures they exert on each other (Milošević, et al. 2018).

Defining the relevant market in the context of digital ecosystems presents a significant challenge for competition analysis. Traditional approaches to market definition rely on clear boundaries based on product substitutability and geographical limitations. However, digital ecosystems often transcend these conventional boundaries by offering a diverse range of interconnected products and services across multiple industries. The multi-sided nature of these platforms, where they serve different groups of users simultaneously (such as consumers, advertisers, and developers), complicates the identification of direct competitors and the assessment of market power. This interconnectedness makes it difficult to delineate a specific market for antitrust purposes, as the competitive constraints may not be adequately captured by traditional market segmentation.

Moreover, digital ecosystems benefit from strong network effects and data-driven advantages that amplify their influence beyond traditional market metrics. The accumulation and utilization of vast amounts of user data enable platforms to improve service quality, personalize user experiences, and create high switching costs, which can entrench their market positions. The convergence of services—such as combining social networking, e-commerce, and media streaming within a single platform—further blurs the lines between distinct markets. This convergence challenges regulators to redefine market boundaries in a way that reflects the dynamic and fluid nature of digital ecosystems. Therefore, competition authorities must adopt more flexible and nuanced approaches to market definition, considering factors like cross-market interactions, network effects, and the pivotal role of data in enhancing market power within digital ecosystems.

Competition concerns in DEs include several critical aspects. First, platforms can exploit their central position to exercise significant control over market participants, potentially resulting in monopolistic behaviors. Second, due to the network effects inherent in DEs, as the value of the platform increases with each additional user, high entry barriers are created for new entrants. Third, platforms often accumulate vast amounts of data which they can use to consolidate their market position and edge out competitors. Lastly, overcontrol by platforms may slow down innovation if the rewards for participants are not set up right.

Cunningham et al. (2018) explores the concept of killer acquisitions, which suggest that established firms strategically acquire potential competitors to eliminate future competition. This phenomenon has raised concerns regarding its impact on market competition and innovation, leading to discussions on the implications for competition policy and antitrust regulations. By exploring killer acquisitions within DEs, the study enhanced understanding of competitive behavior and strategic decision-making in the digital marketplace.

The negative impact of DEs on market competition is advocated by Koski et al. (2020). The authors investigated the impact of acquisitions by major US-based technology firms on market entry dynamics and venture capital financing across various product markets globally from 2003 to 2018. Their findings suggest that these acquisitions led to decreased market entry rates and reduced venture capital funding indicating a "kill zone effect." Particularly during the 2010s, when technology firms gained more access to user data, this effect intensified. Additionally, acquisitions of platform companies by technology giants were found to decrease market entry in non-platform markets, with a notable decline in available venture capital financing observed in the US compared to the European Union.

The comprehensive theoretical background highlight the complexity of competition within digital ecosystems and underscore the necessity for an understanding of these dynamics for effective competition policy formulation. The current fund of academic knowledge, while extensive, offers some conclusions, suggesting a pressing need for tailored antitrust regulations that can better address the digital realities. As DEs continue to influence global markets deeply, the recalibration of competition policies must be prioritized to foster fair competition and encourage innovation, ensuring the long-term sustainability of digital markets and protecting the interests of all market participants.

In the next part of the paper, we will present the case study analyses with the aim to investigate the nature of competition that can appear between DEs. These case study will illustrate the practical implications of theoretical concepts discussed and provide real-world examples of how Des interact competitively across various sectors. By examining specific instances of competition between famous DEs, the following part of the paper aims to highlight the unique challenges and opportunities that these digital interactions present. Through the following analyses, we intend to offer deeper insights into how digital ecosystems can both foster innovation and pose new challenges for competition policy, thereby guiding more effective regulatory practices and strategies.

3 Case Study Analysis

The case study methodology is widely used in academic research to provide an in-depth, contextual analysis of a specific research topic. This qualitative research method allows researchers to explore complex issues in real-life settings, offering rich insights that quantitative methods may not capture. A case study typically involves gathering data through various sources, such as interviews, observations, and existing scientific publications to enhance the validity of findings. In academic papers, the case study approach is particularly useful when examining new or emerging phenomena where existing theories may not fully apply. Case study methodology is widely used in various research fields such as business (Zardini et al., 2023), project finance (Spasenic et al., 2022a; Spasenic et al., 2022b; Spasenić et al., 2022), education (Hojeij et al., 2023) environment and ecology (Neri et al., 2023; Percy & Edwin, 2023).

The case study analysis explores the complex competition within DEs at three levels: (i) between platforms, (ii) between platforms and their complementors, and (iii) among the complementors themselves. By looking at examples like Apple's iOS versus Google's Android, Microsoft's strategy of integrating other services, and the competition among music streaming services, this analysis shows how these relationships influence the growth, operation, and competition of DEs. DEs involve complex interactions that stretch across many industries. This case study looks at how platforms and their complementors, compete in ways that go beyond traditional industry limits. This competition affects not just their market share but also how these ecosystems grow and innovate. The study uses the term "**moligopoly**" (Petit, 2020), a mix of oligopoly and monopoly competition, to help explain these complex dynamics.

The ubiquity of digital ecosystems has made them essential tools for most people worldwide, largely driven by innovative software applications (apps) that have revolutionized work, leisure, and social connectivity. The global app economy is remarkably substantial, valued at approximately $1.7 trillion, highlighting the significant role these ecosystems play in the modern economy (The App Association, 2020). Apps have become integral not only to commerce and professional activities but also to social networking, democratic participation, citizen science, and entertainment.

This explosive growth has intensified competition between digital ecosystems as they strive to attract users, developers, and partners. Major platforms like Apple's iOS, Google's Android and Microsoft are in constant competition to expand their user bases and enhance their app offerings. The vast and growing app economy creates high stakes for these ecosystems, pushing them to innovate and offer superior services. This competitive environment benefits consumers through improved products and services but also raises important considerations for competition policy. As these ecosystems expand, concerns emerge regarding market dominance, potential barriers to entry for new competitors, and the need for regulations that ensure fair competition while fostering innovation.

3.1 Apple Ecosystem

In Apple's digital ecosystem, the company maintains complete control by supplying the devices (iPhones), the mobile operating system (iOS), and the App Store where users download applications. Apple manufactures its own devices, which consumers can purchase directly from Apple or through authorized third parties. Since all iPhones run exclusively on Apple's proprietary iOS, the company tightly integrates hardware and software. This means Apple can dictate the terms of how apps and services function within its ecosystem, influencing everything from app availability to user experience. This high level of control allows Apple to shape its digital environment, which has significant implications for competition with other digital ecosystems like Google's Android.

Apple's iOS stands as a prominent example of a digital ecosystem with vast influence. As the operating system for all of Apple's mobile devices, iOS not only provides a seamless, user-friendly interface but also serves as a gateway for a multitude of applications and services tailored to enhance user experience. Its integrated environment supports a wide range of functionalities from basic phone operations to advanced computing capabilities that cater to both personal and professional needs. The exclusivity of iOS, being available only on Apple hardware, creates a unique market niche that supports Apple's brand loyalty while also fostering a tightly controlled ecosystem. This control allows Apple to ensure security and functionality but has also been a point of contention in discussions about market competition, as it limits the degree of inter-platform operability.

3.2 Google Ecosystem

In the Google digital ecosystem, Google offers its Android operating system to device manufacturers through licensing agreements. This means that companies like Samsung, LG, and Motorola can use Android on their smartphones and tablets. Alongside this, Google operates its own app store called the Google Play Store, which is the main source for apps on Android devices. Because Android is used by a vast number of devices worldwide, the Google Play Store has become the primary hub for app distribution in this ecosystem. This arrangement significantly impacts the competition between digital ecosystems, as it allows Google to extend its reach across numerous devices and compete directly with other ecosystems like Apple's iOS.

Google's Android OS exemplifies a vast and open digital ecosystem that contrasts sharply with Apple's iOS. As an open-source platform, Android allows a wide range of manufacturers to adapt and use its system, resulting in a diverse array of devices that cater to all segments of the market, from budget to premium. This openness has enabled Android to capture the largest share of the global smartphone market, fostering a rich, competitive environment for app developers and hardware manufacturers alike. Android's flexibility allows for extensive customization and integration, offering users and developers more control over their devices. However, this same openness can lead to fragmentation, with variations in device capabilities and user experiences across different manufacturers and models. This ecosystem moves on its ability to be highly inclusive, yet faces challenges in maintaining uniformity and security across its landscape.

3.3 Competition Between Apple's and Google's Ecosystem

The rivalry between Apple's iOS and Google's Android shows a direct competition between two big mobile platforms. Each platform tries to attract more users and developers by offering unique features, better security, and improved user experiences. This competition leads to constant technological improvements and affects the plans of developers and advertisers who want to connect with the most active and largest groups of users. The mobile app ecosystem serves as a prime example of competition between digital ecosystems, characterized by its complexity and the involvement of various stakeholders and technologies. Central to this ecosystem are two dominant firms responsible for the leading mobile operating systems that power the vast majority of the world's apps: Apple Inc., with its iOS platform, and Google LLC., with Android. These companies control vertically integrated distribution chains, overseeing both the operating systems and the primary channels through which mobile apps are distributed. This vertical integration positions them as significant gatekeepers in the mobile app market, allowing them to influence how apps exist, function, and reach consumers.

Unlike desktop computers, where software can be obtained from multiple sources, mobile devices primarily rely on centralized app stores—specifically, Apple's App Store and the Google Play Store—for the distribution of third-party apps. The policies, technical requirements, and guidelines imposed by these app stores fundamentally shape the mobile app ecosystem. While Apple and Google manage their app ecosystems differently, both exert substantial control over app distribution, which can include restricting certain apps or enforcing specific monetization models. This gatekeeping role has significant implications for competition, as it affects the ability of app developers to access the market and compete effectively. The strategic decisions made by these platforms can influence innovation, consumer choice, and the overall dynamics between developers, consumers, and the platforms themselves, highlighting the intricate competition between digital ecosystems in the digital ecosystems industry.

The mobile app store model, largely controlled by Apple and Google, has undeniably provided numerous benefits to both app developers and users. It has made discovering, downloading, and updating apps straightforward, contributing significantly to the growth of the mobile app industry. However, this model has also created less-than-ideal competitive conditions. The policies enforced by Apple and Google introduce unnecessary barriers and costs for app developers. These obstacles range from high fees for access to restrictions that favor certain apps over others. Consequently, developers may face higher development and rollout costs, limited app features, strained customer relationships, and challenges in reaching a wide audience.

Moreover, the structure of these digital ecosystems allows Apple and Google to extract a disproportionate share of the economic gains. They often receive a larger portion of the profits compared to other participants in the ecosystem, like app developers. The incentive structures created by their current policies make it difficult for alternative distribution models, such as other app stores or web browsers, to compete effectively (Nadler & Cicilline 2022). These alternatives often operate at a significant disadvantage with limited functionality, making it challenging for them to establish a foothold in the market.

This scenario means that app developers and alternative app stores are not competing on an even playing field. They face hurdles not only in competing with each other but also against the products and services offered by Apple and Google themselves, who act as gatekeepers. For consumers, this can lead to several negative outcomes: inflated prices due to fees collected by the gatekeepers, stifled innovation because of policy decisions that limit access to certain smartphone features, and a reduced choice of apps, as some are not featured or even accessible on their devices. In essence, while the current mobile app store model has fueled the growth of the app economy, it also raises significant concerns about competition within digital ecosystems. Addressing these issues is crucial to ensure that app developers can innovate freely, alternative distribution methods can emerge, and consumers can enjoy a wider range of choices and fair pricing.

The competition between Apple's iOS and Google's Android illustrates two distinct approaches to ecosystem management. Apple's closed, integrated ecosystem contrasts with Google's open model, highlighting the tension between control and innovation. These differences underscore the need for nuanced competition policies that address platform power and entry barriers in digital markets.

3.4 Microsoft Ecosystem

Microsoft's strategy of integrating a range of services into its platforms exemplifies its approach to creating a comprehensive digital ecosystem. By embedding services such as the Edge browser, Microsoft Stream for media streaming, and Microsoft Teams for messaging and collaboration directly into its Windows operating system, Microsoft ensures that users have immediate, seamless access to a broad suite of essential tools. This integration enhances productivity and convenience by enabling users to navigate, communicate, and collaborate without ever leaving the Microsoft environment. The strategy not only boosts user engagement and satisfaction by providing a unified experience across multiple functionalities but also strengthens Microsoft's market position. It creates stronger barriers to entry for competitors and encourages user loyalty by offering a cohesive, adaptable and flexible platform that effectively meets a wide range of digital needs. Microsoft's strategy of adding browsers, media streaming, and messaging features into its Windows operating system shows how a platform can compete with its complementors. By integrating these features, Microsoft took over valuable functions that other companies provided, increasing its control over the ecosystem.

3.5 Competition Between Music Apps

Spotify, Deezer, Tidal, and Apple Music represent the leading competitors in the fiercely contested music streaming industry. Each platform strives to distinguish itself through unique features: Spotify offers algorithm-driven music discovery, Deezer provides a vast library with high-fidelity sound, Tidal appeals to audiophiles with its lossless audio quality and strong ties to artists, and Apple Music leverages its integration across Apple devices and exclusive content deals. These services compete not just on music catalogs and streaming quality but also on user interface design, playlist curation, and global market reach. As they vie for dominance, they continually innovate and refine their offerings to enhance user engagement and increase their subscriber base, thereby driving the evolution of how music is consumed and appreciated worldwide.

This situation highlights how platforms can take in innovations from complementors, which might limit outside innovation but strengthen the platform's own market position. The competition between music streaming services like Spotify, Deezer, Tidal, and Apple Music shows how companies that offer similar services can compete within the same ecosystem. Each service tries to stand out by offering large music collections, exclusive content, and customized experiences for users. This competition not only affects what consumers choose but also encourages improvements in technology and services as each platform aims to attract more users and increase their spending.

3.6 Results of the Case Study Analysis

The competition between digital ecosystems, as exemplified by Apple's iOS versus Google's Android, Microsoft's integration strategies, and the rivalry among music streaming services like Spotify, Deezer, Tidal, and Apple Music, reveals multifaceted dynamics that shape the digital market landscape.

Firstly, the platform vs. platform competition between Apple and Google showcases how two dominant ecosystems can coexist while offering distinct value propositions. Apple's closed ecosystem emphasizes tight integration and control over both hardware and software, resulting in a consistent user experience but limiting third-party customization. Conversely, Google's open-source approach with Android allows for widespread adoption by various device manufacturers, fostering innovation and diversity but also leading to fragmentation and security challenges.

Secondly, the platform vs. complementor dynamic is evident in Microsoft's strategy of integrating browsers, media streaming, and messaging services into its Windows operating system. By incorporating these functionalities, Microsoft extends its control over the ecosystem, potentially marginalizing independent developers. This strategy can stifle innovation among complementors, as the platform's built-in services compete directly with third-party offerings, reducing the incentive for external innovation.

Thirdly, competition among complementors is illustrated by the music streaming industry, where services like Spotify, Deezer, Tidal, and Apple Music vie for market dominance. Each service differentiates itself through exclusive content, user interface design, audio quality, and personalized user experiences. This intense competition drives technological advancements and benefits consumers through improved services but also leads to market saturation and high customer acquisition costs.

The close connection between different levels of competition is important for understanding how they impact the health and development of ecosystems. The 'moligopoly' competition framework shows how these types of competition are linked, with actions at one level influencing results at another. For example, strong tactics by platforms might decrease the variety of complementors, while strong competition among complementors can boost innovation but might also result in fewer companies dominating the market.

Considering the implications for competition policy, policymakers must consider the complex competition dynamics when designing regulations for digital markets. Traditional antitrust approaches may not fully address the competition interactions in DEs. Therefore, a customized approach that addresses the specifics of 'moligopoly' competition is crucial for promoting healthy competition and preventing anti-competitive practices in DEs. This case study analysis underscores the complexity of competition in DEs across multiple levels. Understanding these dynamics is critical for stakeholders, including regulators, companies, and developers, to navigate and succeed in these environments. Future policies and business strategies should reflect the realities of ecosystem-based competition to enhance innovation, consumer choice, and market health.

4 Recommendations for Increasing Competition and Innovation

To foster a more competitive and **innovative environment** within digital ecosystems, it is essential to promote interoperability and the adoption of open standards. Encouraging platforms to utilize interoperable technologies allows for seamless integration and compatibility across different ecosystems. This approach reduces barriers to entry for new competitors and enables complementors to develop services that operate on multiple platforms, enhancing consumer choice and stimulating innovation.

Enhancing regulatory oversight on platform practices is also crucial. Implementing stronger regulatory frameworks can monitor and prevent anti-competitive behaviors by dominant platforms, such as self-preferencing or imposing unfair restrictions on third-party developers. Increased oversight ensures a level playing field, enabling smaller competitors to compete effectively and fostering a healthy competitive landscape.

Facilitating data portability and giving users greater control over their data is another important step. Mandating data portability standards empowers consumers to easily transfer their data between services and platforms. This reduces lock-in effects and encourages platforms to compete based on the quality of their services rather than relying on user captivity, ultimately benefiting consumers and promoting fair competition.

Supporting alternative distribution channels can significantly enhance competition within digital ecosystems. Encouraging the development and acceptance of alternative app stores and distribution methods beyond the dominant platforms' own channels diversifies the marketplace. This reduces the gatekeeping power of major platforms, provides developers with more options, and stimulates competition in app availability and pricing, offering consumers greater choice.

Implementing fair revenue-sharing models is vital for incentivizing innovation. Advocating for transparent and equitable financial arrangements between platforms and complementors ensures that developers are fairly compensated for their contributions. This motivates developers to create innovative products and services, enriching the ecosystem and providing more value to consumers.

Preventing anticompetitive mergers and acquisitions is necessary to maintain a competitive market. Scrutinizing mergers and acquisitions that could reduce competition, such as "killer acquisitions" where dominant firms acquire emerging competitors to eliminate potential threats, helps preserve market plurality. This prevention of monopolistic control encourages ongoing innovation from a diverse range of companies.

Encouraging collaborative innovation between platforms and complementors can accelerate technological advancements. By fostering environments where platforms actively collaborate with complementors through developer programs and joint innovation initiatives, ecosystems can be enriched with a variety of high-quality services and products. This collaboration not only benefits the platforms and developers but also enhances the user experience.

Increasing transparency in platform policies is important for building trust among all stakeholders. Requiring platforms to be open about their policies, algorithms, and decision-making processes allows for informed decision-making by developers and consumers alike. Transparency prevents unfair competitive advantages derived from obscure practices and promotes fairness within the ecosystem.

Educating consumers and promoting digital literacy can also influence market dynamics positively. Investing in consumer education programs enhances understanding of digital ecosystems, data privacy, and the implications of platform choices. Informed consumers are empowered to make choices that promote competition, such as opting for services that prioritize privacy or innovation, thereby influencing market behavior.

Finally, adapting competition policies to reflect the realities of the digital marketplace is essential. Updating antitrust laws and competition policies to consider the unique characteristics of digital ecosystems such as network effects, the centrality of data, and rapid innovation cycles, ensures that regulations remain effective. Modernized policies are better equipped to promote competition and curb anti-competitive practices in the evolving digital environment.

By implementing these recommendations, stakeholders can enhance competition and innovation within digital ecosystems. This approach not only benefits consumers through improved services and greater choice but also stimulates economic growth by encouraging new entrants and fostering a dynamic, innovative environment. The goal is to create a balanced ecosystem where platforms, complementors, and consumers all thrive through fair competition and collaborative progress.

5 Conclusion

This paper has provided a detailed exploration of the competition dynamics within DES, emphasizing the relationships at multiple levels. The extensive literature review and solid theoretical framework used in this study have provided a deep understanding of these competition environment, which is crucial for anyone involved in the tech industry such as DE, from business strategists to policymakers. The concept of "moligopoly," discussed in this paper, offers a new perspective on these interactions, showing that competition in DE involves more than just fighting for market share.

The examples of Apple's iOS versus Google's Android, Microsoft's strategy of integrating features, and the rivalry among music streaming platforms like Spotify and Apple Music illustrate how these ecosystems are constantly evolving and influencing technological advancements and market dynamics. When considering the rules for competition policy, it's important for policymakers to recognize the complex ways that companies compete in digital markets. Traditional rules might not fully capture the unique interactions in DEs. Therefore, a tailored approach that specifically addresses the unique type of competition called 'moligopoly' is essential to keep competition fair and prevent unfair practices in DEs. This case study highlights how complicated the competition is in DEs, occurring at many different levels. It's important for everyone involved, from government regulators to businesses and software developers, to understand these interactions. Future rules and business plans should consider how competition works in these ecosystems to encourage innovation, offer more choices to consumers, and maintain a healthy market.

Ensuring fair competition on a level playing field is crucial for generating more choices, better quality, and lower prices for consumers. A fair, open, and competitive marketplace has long been a cornerstone of the global economy. As digital ecosystems become increasingly integral to economic growth, it is imperative to promote competition and innovation within this space. When companies compete under equitable conditions, it fosters innovation and allows the natural competitive process to determine market leaders, ultimately benefiting consumers and the economy as a whole.

Understanding the growing importance of digital ecosystems highlights the need for robust competition policies and antitrust laws tailored to these unique markets. Antitrust laws play a vital role in combating anticompetitive conduct and serve as essential tools within the mobile app ecosystem and beyond. By preventing monopolistic behaviors and ensuring that no single entity can unfairly dominate the market, these laws help maintain a dynamic and innovative digital environment where new entrants could succeed.

Fostering a fair and competitive environment within digital ecosystems is vital for encouraging innovation, offering consumers better products and services, and ensuring the overall health of the economy. Policymakers, regulators, and industry stakeholders must collaborate to adapt competition policies to the challenges of the digital age. By doing so, they can ensure that digital ecosystems remain open, competitive, and beneficial for all participants, thus sustaining the foundational principles of a thriving global economy.

In summary, the insights from this research emphasize the need for well-informed strategies and policies to manage competition in digital markets. As these markets grow and change, understanding their complex competitive structures will be key to promoting innovation and fair competition. This paper contributes to the broader conversation on digital competition, suggesting directions for future research and policy development that keep pace with the rapid changes in the digital economy.

This study highlights the evolving nature of competition within digital ecosystems and the need for adaptive policy measures. Future research could further explore how these policy changes can be implemented across different jurisdictions and what impact they would have on innovation in digital markets.

References

Barjaktarović, R.S., Milošević, S., Rakić, N.: Digital banking services from customer perspective. In: Sustainable Business Management and Digital Transformation: Challenges and Opportunities in the Post-COVID Era, pp. 210–227. Springer International Publishing, Cham (2022)

CMA. Online Platforms and Digital Advertising: Market Study and Interim Report. https://assets.publishing.service.gov.uk/media/5fa557668fa8f5788db46efc/Final_report_Digital_ALT_TEXT.pdf (2020). Accessed 21 May 2024

Cunningham, C., Ederer, F., & Ma, S.:. Killer acquisitions. SSRN Electron. J. (2018), Vol. 129, Issue 3, pp. 649–702 doi:https://doi.org/10.2139/ssrn.3241707. Accessed 21 May 2024

Filistrucchi, L., Géradin, D., van Damme, E. Affeldt, P.: Market definition in two-sided markets: Theory and practice. J. Compet. Law Econ., 10(2), 293–339. (2014), doi:https://doi.org/10.1093/joclec/nhu007, Accessed 21 May 2024

Hylton, K.N.: Digital platforms and antitrust law. SSRN Electron. J. (2019), doi:https://doi.org/10.2139/ssrn.3381803. Accessed 25 May 2024

Hein, A., et al.: Digital platform ecosystems. Electron. Mark., 30(1) (2019), 87–98, doi:https://doi.org/10.1007/s12525-019-00377-4. Accessed 21 May 2024

Hojeij, Z., Baroudi, S., Meda, L.: Preservice teachers' experiences with classroom management in the virtual class: A case study approach. Front. Educ., 8. (2023), doi:https://doi.org/10.3389/feduc.2023.1135763. Accessed 21 May 2024

Ilchenko, S.: Research of the factors and conditions for development of the digital economy in Russia. SHS Web Conf., 101, 02010, (2021), doi:https://doi.org/10.1051/shsconf/202110102010. Accessed 27 May 2024

Iossa, E. Rey, P. Waterson, M.: Organising competition for the market. J. Eur. Econ. Assoc. Volume 20, Issue 2, 2022, 822–868. (2021), doi:https://doi.org/10.1093/jeea/jvab044, Accessed 07 June 2024

Jacobides, M. and Lianos, I.: Ecosystems and competition law in theory and practice. Ind. Corp. Chang., 30(5), 1199–1229. (2021), doi:https://doi.org/10.1093/icc/dtab061. Accessed 22 May 2024

Kapoor, R.: Ecosystems: Broadening the locus of value creation. J. Organ. Des., 7(1). (2018), doi:https://doi.org/10.1186/s41469-018-0035-4, Accessed 23 May 2024

Koski, H., Kässi, O. and Braesemann, F.: Killers on the Road of Emerging Start-ups—Implications for Market Entry and Venture Capital Financing. ETLA Working Papers No 81. (2020), http://pub.etla.fi/ETLA-Working-Papers-81.pdf, Accessed 16 May 2024

Kronblad, C., & Envall Pregmark, J.: Responding to the COVID-19 crisis: The rapid turn toward digital business models. J. Sci. Technol. Policy Manag., 15(3), 451–467. (2021), doi:https://doi.org/10.1108/jstpm-10-2020-0155, Accessed 22 Aug 2024

Lianos, I.: Value extraction and institutions in digital capitalism: Towards a law and political economy synthesis for competition law. Eur. Law Open, 1(4), 852–890. (2022), doi:https://doi.org/10.1017/elo.2023.2. Accessed 10 Aug 2024

Lipovenko, M., Gostilovich, A., Gostilovich, S., Ivanov, K., & Liu, M.: Patterns and principles of the development of DEs. Nexo Rev. Cient., 35(1), 174–185. (2022), doi:https://doi.org/10.5377/nexo.v35i01.13929. Accessed 17 Aug 2024

Lončar, D., Milošević, S.: Characteristics of anti-monopoly policy and the effects of its application in Serbia. Ekonomika preduzeća, 115–128 (2013, 2013)

Martens, B.: Data access, consumer interests and social welfare: An economic perspective. SSRN Electron. J. (2020), https://ssrn.com/abstract=3605383. Accessed 24 Aug 2024

Milošević, S., Popović Markopulos, J., Grahovac, J., Ravić, A.: Pricing benchmark in market definition: Theoretical background and practical application. In: Competition Authorities in South Eastern Europe: Building Institutions in Emerging Markets—Contributions to Economics, pp. 173–189. Springer, Cham (2018)

Milošević, S.: Redefining competition in the digital age. In: Fintech in the Context of the Digital Economy: Opportunities and Challenges, pp. 178–193. Balkan Institute of Science and Innovation—BISI, Belgrade (2023)

Nadler, J., Cicilline, D.N.: Investigation of Competition in Digital Markets, (2022). https://www.govinfo.gov/content/pkg/CPRT-117HPRT47832/pdf/CPRT-117HPRT47832.pdf. 17 Aug 2024

Neri, A., Negri, M., Cagno, E., Kumar, V., Garza-Reyes, J.A.: What digital-enabled dynamic capabilities support the circular economy? A multiple case study approach. Bus. Strategy Environ., 32(7), 5083–5101. (2023), doi:https://doi.org/10.1002/bse.3409, Accessed 27 May 2024

Percy, A.J., Edwin, M.: Feasibility assessment and prioritization of renewable energy resources: Towards a energy transition for the society and the environment—a case study approach. Environ. Dev. Sustain. (2023), doi:https://doi.org/10.1007/s10668-023-03799-5, Accessed 23 May 2024

Petropoulos, G.: Competition Economics of DEs—Note by Georgios Petropoulos. OECD. Directorate for Financial and Enterpirese Affairs Competititon Committeee. (2020), https://one.oecd.org/document/DAF/COMP/WD(2020)91/En/pdf, 17 May 2024

Perić, N., Milošević, S., Pešić, S.: Impact of platforms and digital markets to competition policy development. In: Institutional Changes as Determinants for Economic Growth in the Republic of Serbia, pp. 315–332. Faculty of Economics, University of Kragujevac, Kragujevac (2022)

Rochet, J.C., Tirole, J.: Platform competition in two-sided markets. J. Eur. Econ. Assoc. **1**(4), 990–1029 (2003)

Snowdon, B., Stonehouse, G.: Competitiveness in a globalised world: Michael Porter on the microeconomic foundations of the competitiveness of nations, regions, and firms. J. Int. Bus. Stud. **37**, 163–175 (2006)

Spasenic, Z., Makajic-Nikolic, D., Benkovic, S.: Integrated FTA-risk matrix model for risk analysis of a mini hydropower plant's project finance. Energy Sustain. Dev., 70, 511–523 (2022a), doi:https://doi.org/10.1016/j.egyr.2022.06.065. 27 May 2024

Spasenic, Z., Makajic-Nikolic, D., Benkovic, S.: Risk assessment of financing renewable energy projects: A case study of financing a small hydropower plant project in Serbia. Energy Rep., 8, 8437–8450 (2022b), doi:https://doi.org/10.1016/j.egyr.2022.06.065. 27 May 2024

Spasenić, Ž., Petrović, D., Benković, S. Key success factors of a shopping mall project: Creditors perspective in Serbia. Manag. Sustain. Bus. Manag. Solut. Emerg. Econ. (2022), doi:https://doi.org/10.7595/management.fon.2022.0006, Accessed 27 May 2024

Spulber, D.F.: The economics of markets and platforms. J. Econ. Manag. Strateg. **28**(1), 159–172 (2018)

Suseno, Y., Laurell, C., Sick, N.: Assessing value creation in digital innovation ecosystems: A social media analytics approach. J. Strateg. Inf. Syst. **27**(4), 335–349 (2018)

The App Association, 2020, State of the U.S. App Economy. Washington, D.C.: The App Association., https://actonline.org/wp-content/uploads/APP-Economy-Report-FINAL.pdf. Accessed 22 Aug 2024

Varian, H.R., Farrell, J., Shapiro, C.: The Economics of Information Technology: An Introduction. Cambridge University Press, Cambridge (2004)

Veljanovski, C.: Algorithmic antitrust: A critical overview. In: Ezrachi, A. and Stucke, M.E. (eds.) Algorithmic Antitrust, 39–64. (2022), Cham: Springer doi:https://doi.org/10.1007/978-3-030-85859-9_2 Accessed 22 May 2024

Xie, X., Xiangyang, M., Tao, N., & Yin, S.: The coupling mechanism of the digital innovation ecosystem and value co-creation. Adv. Environ. Eng. Res., 04(1), 1–9. (2023), doi:https://doi.org/10.21926/aeer.2301013 Accessed 25 May 2024

Yu, H. and Liu, J.: Analysis of the market definition of non-intermediary service platform in antitrust law. Stud. Law Justice, 1(1), 1–14 (2022), doi:https://doi.org/10.56397/slj.2022.11.01. Accessed 28 May 2024

Zardini, A., Ceesay, L. B., Rossignoli, C., Mahto, R.: Entrepreneurial business network and dynamic relational capabilities: A case study approach. Int. J. Entrep. Behav. Res., 29(2), 328–353 (2023), doi:https://doi.org/10.1108/ijebr-01-2022-0085. Accessed 26 May 2024

Integrated State Audit: A Comprehensive Approach

Nemanja Jakovljević[1](✉), Veljko Dmitrović[2], and Sonja Ivančević[2]

[1] Grant Thornton d.o.o. Beograd, Belgrade, Serbia
jakovljevic.i.nemanja@gmail.com
[2] University of Belgrade—Faculty of Organizational Sciences, Belgrade, Serbia
veljko.dmitrovic@fon.bg.ac.rs

Abstract. Integrated state audit, as a combination of the financial statements audit, the performance audit, and the information systems audit, offers numerous advantages to state auditors, enabling greater impact without necessarily increasing the number of employees. Integration can help state auditors remain relevant and up to date in fulfilling their dynamic roles as advisors and confidants in a rapidly changing environment, particularly following the COVID-19 pandemic. The strength of integrated auditing is that it shows not only where money is spent, but also how it is spent, which is very important knowledge in the public sector. Citizens of a country, as taxpayers and key stakeholders, want government bodies and state-owned enterprises to spend their money wisely, efficiently, and effectively. This is why **public sector auditors** exist, and integrated auditing can provide this assurance. Such concern is completely different from that of private sector stakeholders, who are focused on making a profit. The aim of the paper is a theoretical consideration of the integrated state audit, by discussing about its stages, possibilities for implementation and advantages. The main conclusion is that the process of integration, combining the financial statements audit, the performance audit, and the information systems audit into a single audit concept that would be applied in supreme audit institutions to individual **audit engagements**, can represent a significant challenge for numerous state auditors, yet the overall benefits can be very substantial.

Keywords: integrated audit · state audit · auditors · engagement flow audit

1 Introduction

The integrated state audit, as a unique combination of financial statement audit, business expediency audit, and information systems audit, offers numerous advantages to state auditors, enabling greater impact without necessarily increasing the number of employees. Integration can help state auditors remain relevant and up to date in fulfilling their dynamic role as advisors and confidants in a rapidly changing environment following the Covid19 pandemic (Jakovljević & Dmitrović, 2023). The strength of integrated auditing is that it shows not only where money is spent but also how it is spent, which should be very important knowledge in the public sector. The citizens of a country, as taxpayers

V. Štavljanin et al. (Eds.): SymOrg 2024, LNNS 1681, pp. 17–26, 2025.
https://doi.org/10.1007/978-3-032-08090-5_2

and key stakeholders, which is why public sector auditors exist, probably want government authorities and state-owned enterprises to spend their money wisely, efficiently, and effectively, and integrated auditing can provide this assurance, which is completely a different kind of concern compared to private sector stakeholders, which are focused on making a profit. Historically, government audits have been associated with obtaining information about an organization's financial transactions and financial reporting. However, audits now include non-financial areas, such as security, information systems, and environmental issues (Zaharco & Petreanu, 2023). There is a growing need for expediency audits in government agencies and state-owned enterprises, and as a result, **supreme audit institutions** that have recognized global trends have hired audit practitioners who specialize in security audits and information systems audits (Hegazy et al., 2022). **Risk assessment** decisions require an increased focus on auditors to broaden their perspectives and think outside the box.

A more holistic view of an integrated audit requires the auditor to modify his perspective and think outside the traditional scope of the audit (Amdanata & Mansor, 2020). A key role in the success of an integrated audit is played by the chief auditor due to higher requirements in terms of qualifications and ability to manage the audit project (Sembiyeva et al., 2024). The lead auditor should have a full understanding of the potential risk of the audit activity in all aspects of the scope of the audit (Flasher et al., 2022). The lead auditor will be required to have sufficient soft skills to ensure effective teamwork between audit staff and others who will provide expertise to the engagement (Jakovljevic, 2021). When planning an integrated audit, the initial step is always to define the topic for the business feasibility audit, which includes the subject of the audit necessary resources, and more (Wellalage et al., 2023). Determining the entities for the audit of expediency, the entities for the audit of financial statements, and the list of information systems they use, which should be reviewed, together with the applied techniques and processes of information security and IT controls, are automatically obtained (Candra et al., 2022). Thus, as part of one audit engagement, a package of individual audit reports would be issued in the areas of financial statement audits, business expediency audits, and information systems audits, which explain individual positions or areas in more detail, and an integrated audit report would be issued with them, which would summarize audit findings within the examined topic and which would provide integrated recommendations for the systematic improvement of deficiencies from the topic of overall importance for a society, that is, for a country (Huynh et al., 2022). The subject of the paper is the analysis of the concept of integrated state audit through the presentation of its elements and the flow of engagement.

2 Planning the Integrated State Audit

Information regarding the processes, systems, and issues in that area should be gathered and examined to comprehend the audit topic and the issues in the audit area (Jeremić et al., 2021). The audit team should create summaries of the actions that were audited as well as first conclusions about the operational issues that were encountered, their root causes, and, if applicable, their effects. Auditors ought to search for recurring issues or themes that are mentioned or supported by various sources (Jakovljevic & Dmitrovic,

2024). The easiest way to convey observations is usually to arrange them into broad areas of preliminary findings, identifying the sources of different observations and providing clues about potential operational issues (Flasher et al., 2022). Knowing how various issues or signs of issues in that area relate to one another is also crucial. Initial results should be adequately described in terms of performance concerns and the information's underlying sources. However, there are no hard and fast rules on the type of data that must be provided or the standards that must be met for the initial evaluation at this planning stage. To guarantee that the conclusions of the primary study may be based on appropriate audit evidence, the auditors should, however, define the right criteria as well as the sources and pertinent types of data for the questions to be addressed in the study (Jakovljević & Jakovljević, 2021). The team must create a work plan for the primary research after discussing the audit problem with management in the preliminary stages. Top management should review and approve a draft audit engagement plan. The work plan's goal is to organize and specify how the audit problem review will be carried out during the primary investigation. The team will be guided through the audit process by the work plan, which will assist them in organizing their objectives and providing analytical justifications. To conduct a high-quality audit that adheres to good governance principles and is timely, inexpensive, and efficient, auditors should design the audit in a way that supports these goals (Huang et al., 2024) (Fig. 1).

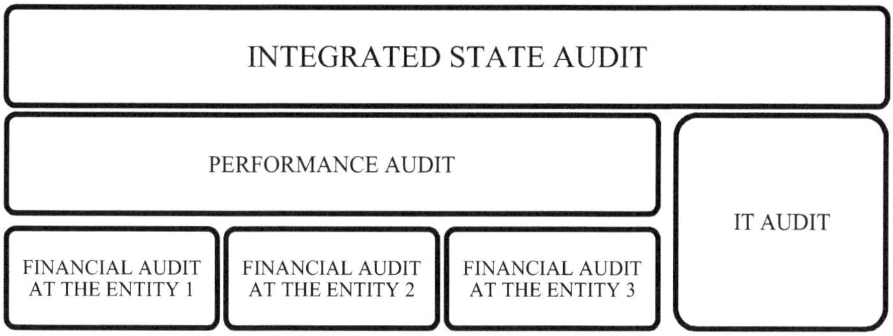

Fig. 1. Integrated state audit, source: Authors

When senior management selects an audit topic that is unconnected to signs or perceptions of issues in a specific area, the audit problem selection method might not be applicable. In these kinds of situations, auditors can employ a results-oriented approach to determine whether specific goals are accomplished, or a system-oriented approach to see whether a certain system is operating as planned. Auditors should use previous study observations and performance problem indicators as a starting point for their investigation of alternative audit difficulties (Efimova et al., 2021). Occasionally, a subject selected by upper management can be sufficiently detailed to serve as an audit challenge (Candra et al., 2022). Under such circumstances, auditors ought to seek out observations that either validate or invalidate the initial notions' plausibility. In addition, auditors want to think about whether there are any other audit problems in this field that would be a better option than the issue that management selected (Viana et al., 2022). The goal of the integrated audit topic selection process should be to optimize the anticipated audit impact while accounting for audit capacity, particularly in terms of professional and human resources, and the degree to which financial, business, and information systems audits can be integrated.

3 Conducting the Integrated State Audit

When it comes to the process of putting an integrated state audit into practice, all entities that are being audited can offer feedback on the draft audit criteria in writing, at a special meeting, or during the introduction meeting. Meetings should be recorded, especially if there are remarks and discussions about the criteria. If required, the audit team should independently modify the audit criteria to reflect those changes. It is critical to work on a shared understanding of the audit. To the extent deemed relevant, further auditee opinions on the audit design and plan should also be taken into consideration. Following the initial meeting, the audit team, the assigned contact person, and other pertinent officials may get together to talk about more operational issues of the audit. The contact person is supposed to help the audit team with contacts and information needed for the audit, including relevant documents, and help set up interviews, among other things. On the other hand, the contact person must be informed of the audit's progress by the audit team leader. Prior to the audit team visiting the auditee's regional or local offices, they must speak with the contact person. Selecting an appropriate data-collecting method is essential to an integrated state audit since it allows auditors to verify information from written documents or field studies in specific circumstances, as well as to obtain enough expertise about the subject from the parties concerned (Melnyk & Shot, 2021).

To obtain a balanced assessment of the situation, interviews with operational personnel, managers at various levels, and other stakeholders should typically be done. Interviews must be well organized and structured, and the one conducting them ought to be knowledgeable about the subjects and topics to be covered. The interview's format is determined by its goal (Dionisijev & Bozhinovska Lazarevska, 2022). The purpose of the integrated state audit team development is to discover and partially verify the existence of flaws, issues, and unfulfilled goals. As such, the inquiries may be more open-ended and wider in nature. However, when the primary study is carried out, the emphasis will change to examine the gaps more thoroughly and determine their causes as well as potential remedies. Typically, an interview happens during a meeting. When the questions are common and relatively easy, telephone interviews might occasionally offer a more cost- and time-effective option. The questionnaire can be used on a sample of population units and offers a methodical way to get data from a specified population. Questionnaires are mostly used to gather information that cannot be found in any document and are crucial for providing context for the results. They can also be used to gather viewpoints (Abutabenjeh & Rendon, 2023). It is crucial that the questions are clearly stated and simple to comprehend. Before the finalized version is distributed to every responder, it is strongly advised that a draft of the questionnaire be delivered to the pilot respondents. For your pilot, use genuine test subjects and specialists. Simultaneously, the audit team must consider a variety of document types, such as regulatory framework and internal policy statements, audit entity budget reports, management reports, applied information systems reviews and documents, file reviews, financial reports, and more. Ministerial instructions, delegating decisions, operational guidelines, manuals, and policy papers should all be gathered by the audit team (Song et al., 2015).

Auditors should also consider modifications made to laws and other records, such as speeches, press announcements, lawsuits, appeals, and case histories, that were instrumental in requiring the modification. Audit entity budget reports include a summary of the entity and include financial data together with details about the objectives and performance of the entity. Organizations usually draft several internal papers outlining current problems and/or recommended lines of action for top management. Such documents should be found and examined by auditors. Interviews and a review of the minutes of management meetings are two ways to find reports. An inspection of the files is a useful source of proof (Huy, 2015). When gathering evidence, auditors should get a list of files from the register of audited subjects to have a general idea of what files are already out there. Working papers provide an organized and concise summary of information file reviews in the auditor's own words. It is imperative that the audit team comprehends that the registered files may not contain all pertinent papers (Bui et al., 2023).

The team might be unaware of further pertinent papers. Auditors should review earlier assessments and reports that are available in the audit area before beginning an audit. Occasionally, their data can be utilized for an audit. It is important to use these supplementary data carefully (Reynolds, 2018). It is advisable to refer to scholarly studies and additional publications that are pertinent to the auditing domain (Dionisijev & Tocev, 2023). While this kind of data might not be sufficient on its own to produce audit evidence, it might offer the theoretical foundation, or points of reference needed to produce more audit evidence. An increasingly significant information source is the Internet. Since more and more government agencies are posting their documents and data online, it is now simple to search through many of the data sources via the Internet (Yaroshchuk et al., 2022). The audit team can gain insight into the reality underlying inspection reports and other official documents through observation. It can offer a more lucid bottom-up perspective of the main problems, which can be contrasted with the image presented in official reports or at the organization's executive level. One of the fundamental risks of people-watching is that the presence of auditors may cause people to behave differently, which will reduce the validity of the data gathered (Grytsyshen et al., 2020). Physical object inspection can yield a wealth of information. The camera is helpful for recording and taking pictures of inspections (Datsii et al., 2021). The primary goal of an inspection is to closely examine a small number of things to provide specific instances of what is seen. It should be verified when the physical state observation is essential to achieving the audit's goals. This can be accomplished by having an observation from two or more auditors, ideally with an auditee representative present as well. To create audit proof, it is also possible to conduct methodical, documented physical inspections of several sites. Rational arguments, computations, comparisons, and dissection of data are examples of analytical evidence (Song et al., 2015). Auditors create fresh evidence that is pertinent to the audit objectives and audit questions by combining data from several sources and conducting both quantitative and qualitative analyses.

4 Integrated State Audit Reporting

Every integrated state audit has a written report that shares the findings with the government, the Parliament, the audit subject, and any other interested parties. The reports that have been published serve as the basis for its actors' assessments of the integrated state audit's efficacy. It is imperative to carefully review the report's veracity, reasoning, and clarity. Writing reports is mostly an analytical process, and an ongoing report-writing process will make auditing easier. The preparation of a report summary should come first when producing a report. The specific evidence that the audit team has obtained in order to achieve the audit goals is referred to as findings. To address the audit questions and fulfill the audit objective, conclusions are developed. The criteria state that every audit finding needs to be backed up by adequate and relevant evidence. Findings must be supported by logical reasoning and pertinent facts at the same time. Before making any decisions, one must consider the surrounding circumstances, apposite reasons, advantages and disadvantages, and many points of view (Dionisijev & Tocev, 2023).

It is important to balance the demand for accuracy against what is practical, affordable, and useful. Perspective and a clear conclusion regarding audit matters should be included in the audit findings. It is important to make sure that the audit's aim, questions, findings, and conclusions are all consistent. The results of the audit should address the audit questions and be in line with the audit's objectives and conclusions. The factual scenario and the audit criteria are compared to produce the audit findings. Discrepancies are conclusions that can indicate possible non-conformances. When there is no disagreement, the auditee has essentially fulfilled the criteria-based expectations. Determining the significance of the discovery and its underlying reasons is crucial for the audit team to accomplish (Jakovljević & Jakovljević, 2022a). It is important for auditors to establish their own standards for assessing every root cause, so they may offer various causes as distinct conclusions. The audit's primary issue is, by definition, the results of the causes it examined. Occasionally, it might be suitable to carry out further analyses of the implications of some findings (Jakovljević, 2022b).

During an audit, it is typical for auditors to note several things that might be used as criteria to assess performance. Since the integrated state audit essentially involves direct reporting, auditors must try to document the actual activities of the examined subjects and assess them based on predetermined standards. It is usually preferable to expand the primary findings at a higher level, conclusions that are supported by several deficiencies discovered, rather than presenting shortcomings one by one (Jakovljević & Petrović, 2023). Additionally, readers will find the report more engaging and find it simpler to retain the key points because of its increased material concentration. Rather than just reflecting on the flaws noted in detail, presenting the key findings at a higher level can also serve as the foundation for recommendations that are likely to make a substantial contribution toward resolving issues or weaknesses found by the audit (Jakovljević, 2022a). When conducting financial statement, performance, and information systems audits, audit teams must make every effort to establish a connection between the results and measure the impact. The team can provide examples of the consequences of poor decision-making, a lack of control, or a lack of concern for the service when the effects cannot be measured. Only essential terms should be included in the matrix, even though the report should include a detailed narrative text outlining the findings (Jakovljević et al., 2022).

The findings matrix aids in the clarification of the findings and enables the team and reviewers to assess if each finding is supported by adequate audit evidence and meets the criteria, all while disclosing the sources of the evidence. Audit findings, on the level of the financial statements, do not typically include an overall assessment of the economy, **efficiency**, and effectiveness achieved at the auditee level in performance audit reports. Conclusions can be supported by quantitative data gathered through sampling or scientific procedures. Asaudit data might be compelling, formulating conclusions to address audit problems may involve a substantial amount of judgment and interpretation (Jakovljević & Jakovljević, 2022b). Before making any findings, it is necessary to weigh the context, pertinent reasons, advantages and disadvantages, and various viewpoints. These conclusions should also be directly related to the audit's goal.

5 Conclusion

One of the key organizational advantages of applying the integrated audit concept is efficiency and **effectiveness**. This exposure helps auditors to develop an understanding of the various business processes of their organization, to become more efficient decision-makers and therefore more complete individuals, to a greater extent qualified to be leaders and bearers of audits in the public sector, and above all state audits. Integrated auditing simultaneously provides organizations with a broader reach and depth in auditing, as well as a powerful tool for looking at complex issues. International standards for public sector auditing require data to be reliable, emphasizing the importance of auditing information systems. Despite its advantages, integrated state audit is not easy to implement. Many supreme audit institutions are insufficiently skilled and lack the staff and skill sets, especially for auditing information systems. The widespread presence of risk aversion, as well as a mindset echoing we've always done things one way, why should we change now when we're doing so well, can be a significant limitation to organizational evolution and transition to integrated auditing. Having adequate and competent personnel resources is a primary concern in the process of establishing the application of integrated audit, and it is often followed by establishing effective communication between members of the audit team and accepting the concept of integrated audit. In the current environment, organizations need to use every resource to become current, so it is advantageous to utilize the skills of auditors.

The transition to an integrated audit certainly represents the evolution of the state audit. Many state audit institutions today conduct separate audits of financial statements, audits of the expediency of operations, and audits of information systems, and the auditors, whether it is about expediency, finances, or information systems, are separated from each other, each in their own field, so they often lose contact with each other, which is a key segment in the process of evolving a professional organization. Through an integrated state audit, everyone will be able to gain greater knowledge about the others' fields. It is a unique, very good opportunity for the internal exchange of knowledge and cross-training, not only in the field of technical skills but also in the field of teamwork, mutual communication, interpersonal relations, trust, and organizational culture, primarily in the context of breaking rivalry between state auditors and team members, as well as developing mutual understanding. Persons engaged in audit work in state audit institutions possess different sets of skills and work with different legal requirements and standards. However, they do not work for personal gain, but for the state interest, which should always be equated with the interest of the community. This means that everyone works to achieve the same ultimate goal, which is to determine whether audit subjects fulfill their responsibilities in accordance with legal authorizations, i.e. whether state bodies and state enterprises spend taxpayers' money efficiently and effectively and to the best of their ability. The main conclusion is that integrating the audit of financial statements, business expediency, and information systems into a single audit concept, which will be applied in supreme audit institutions to individual audit engagements, can pose a significant challenge for many state auditors, yet the overall benefits can be substantial.

References

Abutabenjeh, S., Rendon, R.G.: Procurement and contract management deficiencies: Analysis of state audit reports. Int. J. Public Sect. Perform. Manag. **11**(1), 81–107 (2023). https://doi.org/10.1504/IJPSPM.2023.128537

Amdanata, D.D., Mansor, N.: Performance of regional government-linked companies in Indonesia: The influence of state audit board and regulations. Jurnal Pengurusan. **58**, 1–13 (2020). https://doi.org/10.17576/pengurusan-2020-58-07

Bui, T.T.T., Pham, T.H., Duong, Q.C., Duong, D.T.: Current situation of state management on mineral mining in the North Eastern Vietnam and tasks proposal for audit activities. Inżynieria Mineralna. **52**(2), 23–28 (2023). https://doi.org/10.29227/IM-2023-02-09

Candra, L.A., Ruslan, A., Arie, M.: The authority of the State Audit Board in managing state finances as an effort to prevent corruption. Jurnal Hukum Volkgeist. **7**(1), 89–96 (2022). https://doi.org/10.35326/volkgeist.v7i1.2810

Datsii, O., Levchenko, N., Shyshkanova, G., Dmytrenko, R., Abuselidze, G.: State decoupling audit of low-carbon agricultural production. Rural Sustain. Res. **45**, 93–112 (2021). https://doi.org/10.2478/plua-2021-0011

Dionisijev, I., Bozhinovska Lazarevska, Z.: The impact of Covid-19 on the state audit in the Republic of North Macedonia—Lessons from the crisis. Audit Financiar. **20**(166), 320–332 (2022). https://doi.org/10.20869/AUDITF/2022/166/011

Dionisijev, I., Tocev, T.: The transparency of the supreme audit institution in the Republic of North Macedonia through the analysis of official website traffic. Audit Financiar. **21**(2), 331–339 (2023). https://doi.org/10.20869/AUDITF/2023/170/012

Efimova, N.P., Vasilev, A.A., Yaroshchuk, A.B.: Quality management of environmental protection in the activity of supreme audit institutions in various states and the prospects for improving the quality based on industrial and manufacturing engineering. Int. J. Qual. Res. **15**(4), 1179–1196 (2021). https://doi.org/10.24874/IJQR15.04-10

lasher, R., Shirley, S., Higgins, J.P.: The structure of state auditor functions in the fight against corruption. Curr. Issues Audit. **16**(1), A18–A26 (2022). https://doi.org/10.2308/CIIA-2020-044

Grytsyshen, D.O., Svirko, S.V., Yaremchuk, I.M.: Preliminary audit as a guarantee of effective management of public funds and other state-owned property. Економіка, Управління ТаАдміністрування. **2**(92), 42–55 (2020). https://doi.org/10.26642/ema-2020-2(92)-42-55

Hegazy, M., Hegazy, K., Eldeeb, M.: The balanced scorecard: measures that drive performance evaluation in auditing firms. J. Account. Audit. Finance. **37**(4), 902–927 (2022). https://doi.org/10.1177/0148558X20962915

Huang, R., Tan, B., Xie, Q., Han, B.: Multidimensional attribution and governance optimization path of state audit to promote rural revitalization in the information age. Appl. Math. Nonlinear Sci. **9**(1), 1–18 (2024). https://doi.org/10.2478/amns-2024-0288

Huy, P.Q.: The benefit of state audit and significance to enhancing the validity of Vietnamese state budget accounting. J. Mod. Account. Audit. **3**, 143–149 (2015). https://doi.org/10.17265/1548-6583/2015.03.003

Huynh, Q.L., Hoque, M.E., Susanto, P., Watto, W.A., Ashraf, M.: Does financial leverage mediates corporate governance and firm performance? Sustainability. **14**(20), 13545 (2022). https://doi.org/10.3390/su142013545

Jakovljevic, N.: Political neutrality in the audit profession: Attitudes of respondents in the Republic of Serbia. BizInfo Blace. **12**(2), 23–38 (2021). https://doi.org/10.5937/bizinfo2102023J

Jakovljević, N.: Analysis of issues of certification of persons in the audit profession in the public sector in the Republic of Serbia. Ekonomski Pogledi. **24**(1), 229–247 (2022a). https://doi.org/10.5937/ep24-37934

Jakovljević, N.: Remote work in audit in the private and public sector in the Republic of Serbia. BizInfo J. Econ. Manag. Inform. **13**(1), 65–70 (2022b). https://doi.org/10.5937/bizinfo2201065J

Jakovljević, N., Dmitrović, V.: Auditor's perception of stress during the Covid 19 in the Republic of Serbia. In: Mihić, M., Jednak, S., Savić, G. (eds.) Sustainable Business Management and Digital Transformation: Challenges and Opportunities in the Post-COVID Era SymOrg 2022. Lecture Notes in Networks and Systems, vol. 562. Springer, Cham (2023). https://doi.org/10.1007/978-3-031-18645-5_35

Jakovljevic, N., & Dmitrovic, V. (2024, March 20–22). AI and internal audit, reporting transformation [Paper presentation]. In: Conference on Organizational Science Development Green and Digital Transition–Challenge or Opportunity, Portorož, Slovenia: University of Maribor Press doi:https://doi.org/10.18690/um.fov.3.2024.27

Jakovljević, N., Jakovljević, J.: The impact of the Covid-19 global pandemic on the responsibility of auditors. Finansije. **1–6**, 92–113 (2021) https://mfin.gov.rs/aktivnosti/asopis-finansije-broj-1–62021

Jakovljević, N., Jakovljević, J.: Attitudes of auditors in the Republic of Serbia about using of virtual meetings. Ekonomski Pogledi. **24**(2), 97–126 (2022a). https://doi.org/10.5937/ep24-39222

Jakovljević, N., Jakovljević, S.: Contracting account audit on social networks. Trendovi u Poslovanju. **10**(2), 52–64 (2022b). https://doi.org/10.5937/trendpos2202052J

Jakovljević, N., Jeremić, M., Jeremić, N.: Consideration of respondents' attitudes towards the concept of ethics in the audit profession. Univerzitetska Misao. **21**, 115–125 (2022). https://doi.org/10.5937/univmis2221115J

Jakovljević, N., Petrović, M.: The role of supreme audit institutions in achieving sustainable agriculture. Revizor. **26**(104), 39–48 (2023). https://doi.org/10.56362/Rev23104039J

Jeremić, N., Jeremić, M., Jakovljević, N.: Agility of internal audit. Revizor. **24**(95–96), 57–76 (2021). https://doi.org/10.5937/Rev2196057J

Melnyk, L.I., Shot, A.P.: Implementation of European requirements of the state audit into the national practice. Bìznes Inform. **9**(524), 177–182 (2021). https://doi.org/10.32983/2222-4459-2021-9-177-182

Reynolds, D.: Environmentalists Fear EPA's state 'audit' policy will limit protections. Inside EPA's Water Policy Report. **27**(19), 16–17 (2018) https://www.jstor.org/stable/e27056603

Sembiyeva, L., Ismailova, A., Christauskas, C., Nurmaganbetova, B.: Enhancement of performance audit for effective budget investments. Public Policy Adm. **23**(1), 39–50 (2024)

Song, C., Tian, Y., Zhang, Y.: Reflection on several important issues of state audit. Contemp. Financ. Econ. **0**(1), 1659 (2015) http://cfejxufe.magtech.com.cn/ddcj/EN/

Wellalage, N.H., De Zoysa, A., Ma, S.: Corporate cash holdings and firm performance in India: An empirical investigation of the effects of audit quality and firm growth. Am. Bus. Rev. **26**(1), 7 (2023). https://doi.org/10.37625/abr.26.1.122-147

Yaroshchuk, A.B., Guliev, A.Y., Mikhaylenko, M.N.: Quality management of higher education within the framework of the socio-investment model of economic growth: State audit and financial control. Front. Educ. **7**, 1057005 (2022). https://doi.org/10.3389/feduc.2022.1057005

Zaharco, S., Petreanu, E.: Performance audit regarding the management of agricultural subsidy funds. Sci. Pap. Ser. Manag. Econ. Eng. Agric. Rural Dev. **23**(4), 951–961 (2023)

Review of Empirical Studies on FinTech

Željko Spasenić[(✉)], Nemanja Milanovic, and Milos Milosavljevic

Faculty of Organizational Sciences, University of Belgrade, Jove Ilića 154, Belgrade 11000, Serbia
zeljko.spasenic@fon.bg.ac.rs

Abstract. Financial technologies have been reshaping the financial industry for several decades, leading to an exponential expansion in academic research on FinTech. This study aims to systematically analyze scholarly works on FinTech that are based on primary data collection methods, including questionnaires and interviews. By employing bibliometric techniques, this paper provides a comprehensive overview of the research landscape, including summary statistics on academic output volume, journals and individual authors productivity, geographical dispersion, and patterns of cross-country collaboration. A total of 264 papers were retrieved from Web of Science database and analyzed employing a bibliometric approach. Thus, this study contributes to both analysis of the evolving FinTech literature and to the methodological framework for bibliometric literature reviews.

Keywords: FinTech · primary data collection · questionnaires · interviews · bibliometric analysis

1 Introduction

In recent decades, modern technology has been revolutionizing the **financial services** industry, with innovations disrupting the traditionally conservative landscape. The development of financial technologies has progressed through three key historical stages: emergence, fermentation, and take-off. Consequently, both academic and practical interest in the intersection of finance and technology has surged significantly.

This heightened attention to the field of fintech is often accompanied by more precise definitions and refined methods for measuring various aspects of this rapidly evolving and multifaceted phenomenon. However, for the purposes of this study, the lack of convergence among the empirical investigations conducted on fintech is particularly noteworthy. The findings from these **primary-source studies** appear to diverge rather than converge on key issues, trends, and dynamics within the fintech landscape, suggesting a need for further synthesis and reconciliation of the existing knowledge in this domain.

Thus far, numerous studies on Fintech were primary studies by their nature and employed data collection techniques such as questionnaires and interviews. These approaches, commonly used to gather data from both individuals and organizations, are valuable for capturing qualitative and quantitative insights into various aspects of

V. Štavljanin et al. (Eds.): SymOrg 2024, LNNS 1681, pp. 27–39, 2025.
https://doi.org/10.1007/978-3-032-08090-5_3

FinTech development. The primary studies on FinTech are related to the technology acceptance and user behaviour [1, 2], blockchain technology and dynamic capabilities [3], entrepreneurial ecosystem and financial inclusion [4, 5], banking, FinTech and digital transformation [6–8], specific segments of FinTech ecosystem, such as InsurTech [9], consumer behaviour and digital transformation [10] and dark side of FinTech and related risks [11].

The objective of this paper is to present a comprehensive review of empirical studies on FinTech that utilize primary data sources. Given the significant body of work that addresses FinTech through primary research, this study adopts a bibliometric approach. To the best of our knowledge, no prior research has systematically examined the literature on FinTech with this specific objective.

The paper is structured as follows. Sect. 2 describes the methodology employed for the bibliometric analysis. Sect. 3 presents the key findings from the study. Sect. 4 provides context for the findings and offers implications and recommendations for both academics and practitioners.

2 Methodology

The methodology used in this study adheres to the three-stage methodology framework previously utilized in [12–14]. The initial phase involved the systematic identification of pertinent scholarly articles within the **Web of Science** (WoS) database, which is confirmed as reliable source of publications for literature reviews across various research areas [15].

The search procedure employed a combination of two distinct keyword pairs linked by the Boolean operator OR: "FinTech OR Financial Technolog*" and "Questionnaire* OR Interview*". By utilizing "FinTech OR Financial Technolog*", the search ensured a thorough exploration of research addressing diverse facets of FinTech, including user perceptions, industry trends, and technological advancements. The second keyword pair effectively encompasses studies that investigate the FinTech phenomenon through primary data collection techniques, such as questionnaires and interviews.

This keyword string was applied to the WoS search engine, targeting titles, abstracts, and keywords of indexed publications. Additionally, the search parameters were refined to include specific types of publications (articles, review articles, and conference proceedings) and to restrict the results to publications in English. The language filter eliminated only three publications, thereby affirming the extensive availability of literature in English. The initial search yielded 323 potentially relevant publications covering the period from 2016 to 2024.

In the subsequent phase, a content analysis was performed on the sample retrieved publications include solely primary-source studies related to FinTech. This process involved a thorough screening of titles and abstracts to select publications that pertain exclusively to original primary research on FinTech. The screening of each publication was independently done by all three authors. The final research sample is formed at the consensus meeting between the authors with the goal to omit any mistakes in the refinement process.

As a result, the final research sample comprised 264 publications, covering the period from 2016 to 2024, with the distribution by publication type presented in Table 1. Information regarding the selected publications, including titles, abstracts, keywords, and document types, was extracted from WoS as a text tab-delimited file for further analysis using VOSviewer software.

Table 1. Sample structure according to the document type

No.	Document type	Number of documents	% of total
1	Article	214	81
2	Proceeding paper	44	17
3	Review article	6	2
Total		264	100

Most of the studies are journal articles, comprising 214 documents, which account for 81% of the total sample. Proceedings papers make up 44 documents, representing 17% of the sample, while review articles contribute 6 documents, amounting to 2% of the total. This distribution highlights the dominance of journal articles in FinTech research, with a smaller but significant portion of studies presented in conference proceedings and a limited number of literature review articles.

Finally, we did a bibliometric and content analysis, which consists of several interconnected steps. First, a descriptive analysis of the final sample of identified studies is implemented to assess the chronological evolution of publications within the defined period. Next, an examination of researcher production, funding sources and cross-country cooperation is performed. Following this, the most influential works in the field, determined by citation counts, are identified and a content analysis is undertaken to identify the crucial topics emerging from the most influential publications. The methodology design is presented in Fig. 1.

3 Results

This section responds to the previously stated research questions by systematically analyzing several key bibliometric indicators. First, the frequency of academic publications is analyzed to identify trends in the research output volume over time. In addition to assessing the productivity of journals and individual authors, the study investigates the geographical dispersion of research on FinTech to highlight regions that are strongly contributing to this field. Moreover, patterns of international collaboration are explored to understand the extent of cross-border networking and partnerships in producing and sharing knowledge related to this emerging field. Finally, the main subcategories within FinTech are identified through publications' keyword analysis of the publications, providing a deeper insight into the thematic focus and emerging trends within the literature.

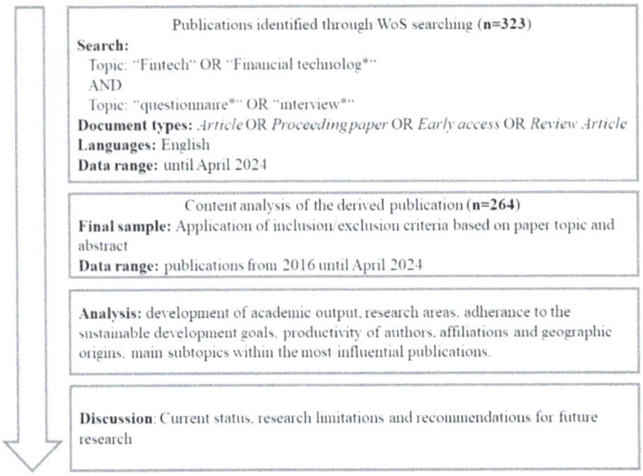

Fig. 1. Methodology design

3.1 Trends in Empirical FinTech Research

Figure 2 highlights a notable rise in empirical studies utilizing primary data on FinTech, reflecting growing academic interest and research activity in the field. The consistent increase in publications from 2016 to 2024 points to the expanding recognition of FinTech's significance across multiple sectors as explained in the following sections. Looking ahead, FinTech research is likely to continue its upward trajectory as scholars explore emerging trends and challenges in this fast-developing sector.

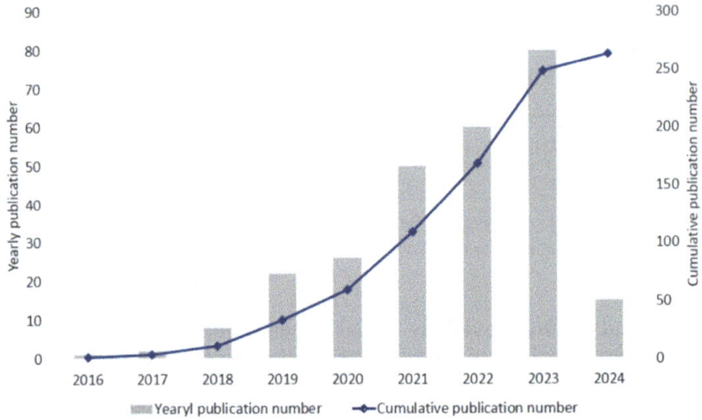

Fig. 2. Growth of Empirical FinTech Research from 2016 to 2024

3.2 Research Areas

The distribution of FinTech-related research papers across various research areas indexed in the WoS demonstrates a strong interdisciplinary focus. According to WoS classification, the same paper might be classified in more than one research area meaning that it contributes to all of them. Most of these studies are concentrated within the business and economics domain, which accounts for 152 papers. This reflects the central role of financial technologies in transforming economic models and business practices.

Following this, computer science emerges as the second most prominent area, with 51 papers, highlighting the technical underpinnings of **FinTech innovations**. Science and technology studies contribute 26 papers, emphasizing the role of technological advancements in shaping the field. Interestingly, environmental science and technology accounts for 20 papers, signifying a growing interest in the intersection of FinTech and sustainability-related topics. Other fields, such as engineering with 18 papers and public administration with 13 papers, show a more specialized focus, exploring the impact of FinTech from both a technological infrastructure and regulatory perspective.

In line with the multi-dimensional nature of FinTech research, the research papers are published in scientific journals spanning across different research areas. The leading journal in FinTech-related primary research is Sustainability, with 12 publications. Close behind is Technological Forecasting and Social Change, with 6 papers, focusing on technological advancements and their social impacts within the FinTech sector. Other notable journals, including Cogent Business Management, International Journal of Bank Marketing, Journal of Financial Services Marketing, and Journal of Science and Technology Policy Management, each contributed 5 papers, reflecting a wide range of academic outlets actively engaging in FinTech research.

3.3 Adherence to the Sustainable Development Goals

The Sustainable Development Goals (SDGs) comprises of 17 global objectives established by the United Nations in 2015, aimed at addressing major global challenges such as poverty, inequality, climate change, environmental degradation, peace, and justice by 2030. These goals provide a framework for promoting sustainable economic, social, and environmental development across all nations. Many scientific journals are strongly encouraging the authors to submit papers that are supporting the achievement one of SDGs.

The papers in research sample align with several different Sustainable Development Goals (SDGs). Notably, 72 papers are linked to the goal of Quality of Education (fourth SDG), reflecting FinTech's role in improving access to education and financial literacy. Industry Innovation and Infrastructure (ninth SDG) is covered by 47 papers, emphasizing FinTech's contribution to advancing technology and infrastructure development.

Additionally, 38 papers focus on No Poverty (first SDG), underlining the importance of FinTech in promoting financial inclusion and providing accessible financial services to low-income populations. The Decent Work and Economic Growth goal (eighth SDG) is addressed by 19 papers, illustrating FinTech's role in creating employment opportunities and fostering economic development. Lastly, 10 papers explore FinTech's potential to Reduce Inequality (tenth SDG), highlighting how it can help bridge financial gaps and improve access to financial resources for marginalized groups. These findings demonstrate FinTech's broad relevance across various SDGs.

3.4 Productivity of Authors

The analysis of author production reveals that no single researcher dominates the field entirely. At the forefront is Alsmadi AA, who has authored 6 papers, followed by Mogaji E and Nguyen NP, who co-authored 4 papers each. Beyond these top contributors, the list becomes more fragmented, with 6 authors who wrote 3 papers and 30 authors contributing 2 papers. The remaining authors in the research sample each published only a single paper.

The diverse research interests of the leading authors highlight once again the multi-disciplinary nature of FinTech research. Their collective focus spans critical areas such as cultural influences, ethical challenges, consumer behavior, and the digital transformation of financial services. This body of work reflects the growing complexity of FinTech and its broad impact across different sectors and markets, particularly in emerging economies.

Alsmadi AA focuses primarily on Islamic FinTech adoption, exploring how religious orientation influences its acceptance [16]. Additionally, Alsmadi's work investigates cryptocurrency adoption and user behavior, examining factors like trust, market uncertainty, and legal environments [17]. His research interest also covers the digital transformation in the banking industry, specifically in areas such as digital accounting [18].

Mogaji E, co-authoring with Nguyen NP, explores the branding strategies of neobanks and traditional banks, emphasizing their adaptation to digital platforms [19]. Their research also delves into the dark side of FinTech, addressing the ethical challenges and risks in mobile money and other financial technologies, especially in emerging markets [20]. Additionally, they also examined consumer interactions with banking chatbots and their implications for financial services [21].

3.5 Author Affiliation, Funding Sources and Geographic Collaboration

The analysis reveals that the most productive institutions in FinTech research are Al Zaytoonah University of Jordan (6 authors), followed by University of Indonesia, University of Oxford, and University of Teesside (each with 5 authors). These affiliations illustrate the global spread of FinTech research, with contributions from both Asia and Europe, highlighting FinTech's international significance and area of interest.

Regarding the geographical origins of the publications, China leads with 33 papers, followed by Indonesia (30), India (28), and the United Kingdom (27), which is the most productive European country. Other significant contributors include Malaysia (21), Jordan (15), United States (13), and Saudi Arabia (11). These figures reflect the dominant role of Asia in FinTech research, with countries like China and India playing pivotal roles.

The described geographical distribution aligns with existing research that confirms Asia's leadership in FinTech adoption rate due to rapid technological advances, favorable regulatory environments, and significant investment. Also, this is in line with the most recent research of Liu and coauthors, who found that China is the highest productive country in fintech research followed by USA and United Kingdom [22].

The funding landscape shows a strong support for FinTech research, particularly from Asia. China's National Natural Science Foundation (NSFC) funded 6 papers, while UK

Research and Innovation (UKRI) supported 4 papers. The Ministry of Research and Technology of Indonesia (RISTEK) and VC Research each funded 3 papers. These funding agencies play a critical role in promoting research and innovation, particularly in emerging markets like China and Indonesia, which are becoming global leaders in FinTech. According to BCG [23], Asia-Pacific region will be the largest fintech market by 2030 while Latin America and will be the fastest growing regions. In line with the development of FINTEC industry it is expected further financial support for academic institutions and research.

The cross-country cooperation among the researchers is presented in Fig. 3. The green cluster, consisting of countries like Bangladesh, Indonesia, Malaysia, China, Taiwan, and Vietnam, plays a crucial role in shaping FinTech research and development in Asia. Partnerships and knowledge sharing within this cluster are essential for the accumulation collective expertise, addressing shared challenges, and driving the sustainable growth of FinTech innovations both regionally and globally.

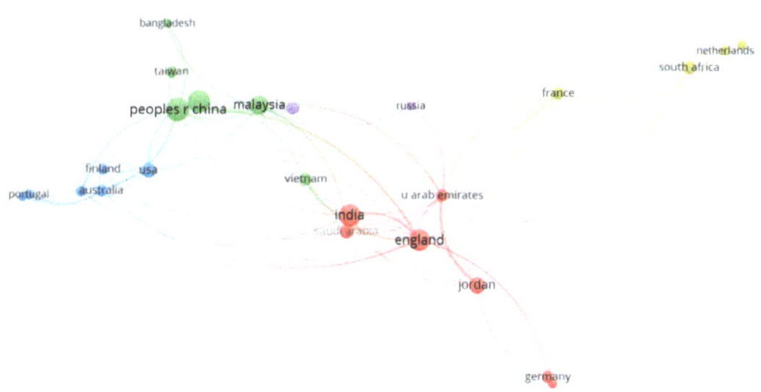

Fig. 3. Cross-country cooperation

The red cluster, comprising India, Saudi Arabia, the UAE, Jordan, and Germany, represents a key network in academic research cooperation within the FinTech field. This group facilitates collaborative efforts across regions, integrating India's digital innovation, Middle Eastern financial leadership, and Germany's regulatory expertise. The cooperation within this cluster strengthens cross-border academic partnerships, advancing FinTech research and development in both the Middle East and Europe.

The blue cluster, comprising the USA, Finland, Australia, and Portugal, represents a significant hub of academic collaboration in FinTech. The author from these countries combines US innovation, European regulatory expertise, and Australia's tech advancements, fostering research cooperation that drives FinTech progress across North America, Europe, and Oceania.

3.6 Main Subtopics within the Most Influential Publications

The analysis of the main subtopics within the most influential publications is based on all keywords with the minimum occurrence of 2 (author keywords and keywords plus generated by WoS). For the sub-sample consisting of the most influential publications, we selected 50 most cited publications. The most cited publication has 205 citations while the 50th on the list has 15 citations.

Using the **VOSviewer software** (version 1.16.17) we identified 6 distinct clusters. The keyword grouping is presented in Fig. 4 where each circle represents a single keyword. The occurrence frequency of keywords is positively correlated with the size of the circles used to represent them; in other words, keywords that appear more frequently are depicted by larger circles. Various colours are utilized to indicate different thematic clusters, which will be examined in the subsequent analysis.

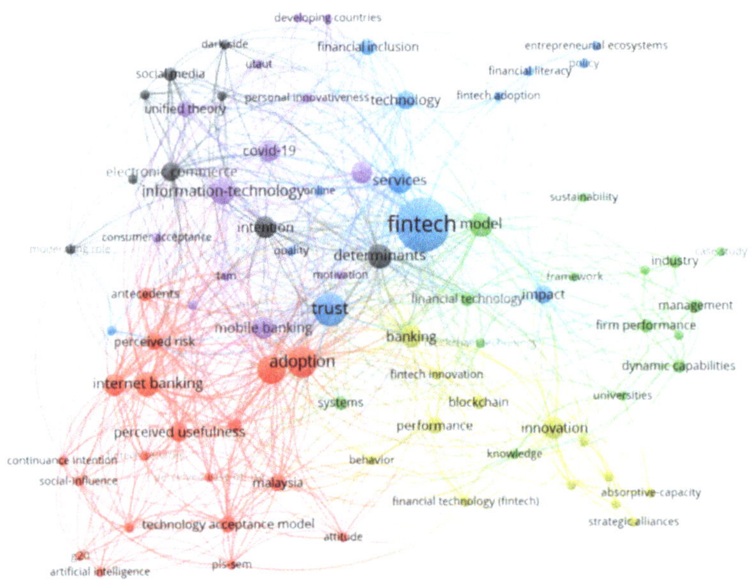

Fig. 4. Thematic clusters within the most influential publications

***Technology Acceptance** and User Behavior Cluster (red cluster).* The Red Cluster centers on technology acceptance and user behavior, particularly in the context of FinTech services such as online banking, e-commerce, and internet banking. It emphasizes factors influencing both initial adoption and long-term engagement, including perceived ease of use, perceived usefulness, attitude, and social influence, which are key components of the Technology Acceptance Model (TAM). Future research could explore how emerging FinTech services, such as decentralized finance and artificial intelligence supported financial platforms, influence user behavior and acceptance models, particularly in diverse cultural and demographic settings.

***Blockchain technology** and dynamic capabilities (green cluster).* This cluster focuses on blockchain technology and its integration into financial technology (FinTech). It highlights the exploration of dynamic capabilities, firm performance, and management in the context of FinTech innovation. The presence of keywords like sustainability and supply chain finance suggests that this cluster addresses how FinTech, especially blockchain, is reshaping industries by improving operational efficiencies and promoting sustainable business models. Further studies could investigate the scalability and interoperability of blockchain technologies across various industries, with a focus on their impact on global supply chain sustainability and cross-border financial transactions.

Entrepreneurial Ecosystem *and* ***Financial Inclusion*** *(blue cluster).* The blue cluster centers on the role of FinTech in enhancing the entrepreneurial ecosystem and promoting financial inclusion. Key terms like financial literacy, FinTech adoption, and policy suggest a focus on creating inclusive financial services through technology. This cluster emphasizes the role of start-ups and entrepreneurial activity in driving financial innovations, while also addressing industry challenges such as the impact of technology on firm performance and innovation policy. Future research should examine how advancements in FinTech can further support underrepresented entrepreneurs and promote financial inclusion in underserved regions.

Banking*, FinTech and* ***Digital Transformation*** *(yellow cluster).* The yellow cluster focuses on the ongoing digital transformation within the banking sector. Key terms such as blockchain, digital transformation, and FinTech innovation indicate that this cluster explores how traditional banking models are being redefined by FinTech innovations. The emphasis on strategic alliances, services, and firm performance points to the collaboration between technology firms and financial institutions in driving this transformation. Additional research is recommended to assess the long-term effects of strategic alliances between traditional banks and FinTech firms, particularly in areas such as data sharing, cybersecurity, and customer-centric innovations.

Consumer Behavior *and Digital Transformation (purple cluster).* This cluster examines consumer acceptance and behavioral intention toward adopting FinTech services, with a focus on developing countries. Keywords such as covid-19, e-loyalty, personal innovativeness, and unified theory suggest that this cluster delves into the psychological and social factors that influence the adoption of digital financial services. It also explores how the pandemic has accelerated digital transformation in financial services and reshaped consumer behavior. Future studies could delve deeper into how psychological traits such as digital trust and anxiety influence consumer adoption of FinTech.

Dark side and risk (grey cluster). The grey cluster highlights the dark side of FinTech and related risks. It examines challenges such as mobile money risks, social media, and consumer behavior in the digital financial space. This cluster focuses on the unintended consequences and ethical concerns associated with FinTech innovations, particularly in emerging markets. The presence of keywords like moderating role, risk, and electronic commerce indicates a critical examination of the potential downsides of rapid technological adoption in financial systems. Further research is needed to develop comprehensive frameworks for mitigating the ethical and security risks associated with FinTech, including regulatory measures for preventing financial fraud and ensuring consumer data protection in digital ecosystems.

4 Conclusion

In recent years, scholarly interest in FinTech phenomenon has grown substantially. The prolific increase in research output within this dynamically growing field has made traditional systematic literature reviews, involving manual examination of large scientific paper datasets, nearly unfeasible. However, the increasing availability of software tools that simplify quantitative analysis has led to a significant surge in the number of bibliometric studies. Their importance is particularly prominent when investigating phenomena related to the application of modern technologies. As noted by Daim and coauthors [24], "bibliometric studies help us understand the past, while also providing a certain degree of foresight into the future application of new technologies."

However, the quality of bibliometric literature reviews heavily relies on the quality of the selected set of papers. To accurately represent the field of interest, the dataset must be as comprehensive as possible. The methodology utilized in this study follows a three-stage approach, facilitating a comprehensive analysis of the 264 publications identified through the search strategy outlined in the Methodology section.

As showed by the results of this study, the academic output on FinTech phenomenon has experienced steady growth in recent years. The findings highlight the distinctly interdisciplinary nature of FinTech research, which extends across a wide range of domains. These include areas such as sustainability, technology forecasting, business and management perspectives, as well as science policy management. This broad scope underscores the multifaceted impact of FinTech and its relevance to both technological innovation and business and economics domain.

The analysis of countries, research institutions, and funding bodies that significantly contribute to FinTech research and development has revealed the leading role of Asian countries. The results confirmed that countries with a well-established financial sector, advanced technological development, and substantial investment in R&D exhibit the highest levels of scientific research output in the field. Furthermore, the proximity to highly developed FinTech ecosystems appears to play a crucial role in enhancing research contributions, underlining the multifaceted nature of this rapidly evolving field. Similarly, the most productive authors of the most influential studies in the field, are predominantly from the same geographic regions, citing leading universities in the Asia-Pacific region as their affiliations, followed by USA and selected EU countries.

The contributions of our study are twofold. Firstly, we contribute to the discussion on bibliometric literature reviews by proposing a three-stage methodological approach for conducting a systematic, comprehensive bibliometric analysis of literature. The methodology is replicable and can be transferred to any other research field. Secondly, by employing quantitative bibliometric analysis techniques, this study addresses the identified research gap by systematically organizing and mapping existing knowledge on the FinTech phenomenon. It identifies key themes and research trends in the field and provides valuable guidance for researchers in designing their future studies by highlighting promising research gaps with significant potential. Additionally, the study offers a valuable tool for managing extensive databases of scientific papers that are impractical to analyze manually.

Although the results of this study and the proposed methodology provide a robust foundation for bibliometric literature reviews, it is important to acknowledge its limitations primarily related to the sampling procedure and the research design. The analysis was primarily based on the metadata of the selected studies, rather than a comprehensive review of the full texts, which represents a limitation of this study that should be considered when interpreting the findings. Also, the selection of keywords such as "interview" and "questionnaire" may have excluded some relevant studies, as these terms are not commonly used as primary descriptors in academic research, thereby introducing a potential bias in the scope of our review.

Given that this is a bibliometric study, it is challenging to avoid certain shortcomings and limitations inherent to this type of research. These include issues related to the selection of publication databases and bibliometric software, the setting of search criteria, and the inevitable subjectivity in drawing conclusions. Although this study primarily employs a quantitative approach, it is also enriched with qualitative interpretations, conclusions, and projections concerning the evolution of FinTech based on available. The integration of quantitative and qualitative aspects in bibliometric studies is often not clearly defined [25], which means that, like all studies of this nature, this research is inevitably subject to some degree of subjective interpretation.

References

1. Milanović, N., Milosavljević, M., Benković, S., Starčević, D., Spasenić, Ž.: An Acceptance Approach for Novel Technologies in Car Insurance. Sustainability. **12**(24), 10331 (2020). https://doi.org/10.3390/su122410331
2. Spasenić Ž., Milosavljević M., Benković S.: The Role of ERP in Digital Transformation: A User Perspective. In proceedings of XVII International symposium Symorg 2020 Business and Artificial Intelligence, online, Serbia, 7–9 September 2020, Belgrade: Faculty of Organizational Sciences (2020)
3. Chang, V., Baudier, P., Zhang, H., Xu, Q., Zhang, J., Arami, M.: How Blockchain can impact financial services – The overview, challenges and recommendations from expert interviewees. Technol. Forecast. Soc. Chang. **158**, 120166 (2020). https://doi.org/10.1016/j.techfore.2020.120166
4. Milosavljevic, M., Milanovic, N., Spasenic, Z., Benkovic, S.: Civic Crowdfunding: Systematic Literature Review and Future Research Agenda. In: the Book of Abstracts of the 2021 International Conference E-Business Technologies "EBT 2021", 10–11 June 2021, pp. 72–74. Belgrade, Serbia (2021)
5. Alaassar, A., Mention, A.-L., Aas, T.H.: Ecosystem dynamics: exploring the interplay within fintech entrepreneurial ecosystems. Small Bus. Econ. **58**(4), 2157–2182 (2021). https://doi.org/10.1007/s11187-021-00505-5
6. Wonglimpiyarat, J.: FinTech banking industry: a systemic approach. Foresight. **19**(6), 590–603 (2017). https://doi.org/10.1108/fs-07-2017-0026
7. Spasenić Ž., Sredojević S., Benković S.: Digital transformation of banking business in Serbia caused by COVID-19 pandemic. In book of abstracts of XVIII International symposium Symorg 2022 Sustainable Business Management and Digital Transformation: Challenges and Opportunities in the Post-COVID era, Serbia, 11–14 June 2022, Belgrade: Faculty of Organizational Sciences, 153–154 (2022a). ISBN 978-86-7680-411-5

8. Spasenić, Ž., Sredojević, S., Benković, S.: Digital transformation of commercial banks in Serbia during COVID-19 pandemic. In: Mihić, M., Jednak, S., Savić, G. (eds.) Sustainable Business Management and Digital Transformation: Challenges and Opportunities in the Post-COVID Era. SymOrg 2022. Lecture Notes in Networks and Systems, pp. 196–209. Springer, Cham (2022b). https://doi.org/10.1007/978-3-031-18645-5_12
9. Milanovic, N., Milosavljevic, M., Zarkic Joksimovic, N.: The emergence of insurtech: A bibliometric survey. In Proceedings of 70th International Scientific Conference on Economic and Social Development–Baku, 25–26 June, 2021, pp. 1124–1133 (2021)
10. Rangaswamy, E., Yong, W.S., Joy, G.V.: The evaluation of challenges and impact of digitalisation on consumers in Singapore. Int. J. Syst. Assur. Eng. Manag. **15**(5), 1704–1716 (2023). https://doi.org/10.1007/s13198-023-02023-x
11. Sampat, B., Mogaji, E., Nguyen, N.P.: The dark side of FinTech in financial services: a qualitative enquiry into FinTech developers' perspective. Int. J. Bank Mark. **42**(1), 38–65 (2023). https://doi.org/10.1108/ijbm-07-2022-0328
12. Spasenic, Z., Milosavljevic, M., Milanovic, N.: Project financing of renewable energy projects a bibliometric analysis and future research agenda. Fresenius Environ. Bull. **31**(8), 7844–7851 (2022)
13. Vukmirovic, V., Spasenić, Ž., Milosavljević, M.: A bibliometric analysis and future research agenda for online labour platforms. Stanovnistvo. **61**(2), 183–207 (2023). https://doi.org/10.59954/stnv.537
14. Milosavljević, M., Spasenić, Ž., Krivokapić, J.: Bibliometric review of participatory budgeting: Current status and future research agenda. Int. J. Financial Stud. **11**(3), 104 (2023). https://doi.org/10.3390/ijfs11030104
15. Li, K., Rollins, J., Yan, E.: Web of Science use in published research and review papers 1997–2017: a selective, dynamic, cross-domain, content-based analysis. Scientometrics. **115**, 1–20 (2018). https://doi.org/10.1007/s11192-017-2622-5
16. Alsmadi, A.A.: Exploring the moderating role of religious orientation on Islamic Fintech adoption. Int. J. Islam. Middle East. Financ. Manag. **17**(2), 310–327 (2024). https://doi.org/10.1108/imefm-09-2023-0315
17. Alsmadi, A.A., Shuhaiber, A., Al-Omoush, K.S.: Rolling the Crypto dice: The interplay of legal environments, market uncertainty, and gambling attitudes on users' behavioral intentions. Hum. Behav. Emerg. Technol. **2023**, 1–17 (2023). https://doi.org/10.1155/2023/8259139
18. Al-Okaily, M., Alsmadi, A.A., Alrawashdeh, N., Al-Okaily, A., Oroud, Y., Al-Gasaymeh, A.S.: The role of digital accounting transformation in the banking industry sector: an integrated model. J. Financ. Report. Account. **22**(2), 308–326 (2023). https://doi.org/10.1108/jfra-04-2023-0214
19. Mogaji, E., Nguyen, N.P.: High street banking on the app: branding strategies of traditionally-driven neobanks. Int. J. Bank Mark. **42**(2), 301–330 (2024). https://doi.org/10.1108/ijbm-12-2022-0529
20. Mogaji, E., Nguyen, N.P.: The dark side of mobile money: Perspectives from an emerging economy. Technol. Forecast. Soc. Chang. **185**, 122045 (2022). https://doi.org/10.1016/j.techfore.2022.122045
21. Mogaji, E., Balakrishnan, J., Nwoba, A.C., Nguyen, N.P.: Emerging-market consumers' interactions with banking chatbots. Telematics Inform. **65**, 101711 (2021). https://doi.org/10.1016/j.tele.2021.101711
22. Liu, Q., Chan, K.-C., Chimhundu, R.: Fintech research: systematic mapping, classification, and future directions. Financ. Innov. **10**(1), 24 (2024). https://doi.org/10.1186/s40854-023-00524-z

23. BCG: Global Fintech 2023: Remaining the Future of Finance (2023). https://web-assets.bcg.com/66/7e/a36d7eab41e2b4b65c3e687a17f5/bcg-qed-global-fintech-report-2023-reimagining-the-future-of-finance-may-2023.pdf
24. Daim, T.U., Rueda, G., Martin, H., Gerdsri, P.: Forecasting emerging technologies: Use of bibliometrics and patent analysis. Technol. Forecast. Soc. Chang. **73**(8), 981–1012 (2006). https://doi.org/10.1016/j.techfore.2006.04.004
25. Donthu, N., Kumar, S., Mukherjee, D., Pandey, N., Lim, W.M.: How to conduct a bibliometric analysis: An overview and guidelines. J. Bus. Res. **133**, 285–296 (2021). https://doi.org/10.1016/j.jbusres.2021.04.070

Mind the Gap: Constraints and Prospects of Neuromarketing and Consumer Neuroscience

Aleksandar Mirčov[1], Tamara Vlastelica[2(✉)], and Jovanka Vukmirović[2]

[1] Sabio Inc., Los Angeles, USA
sasha@sabioholding.com

[2] Faculty of Organizational Sciences, University of Belgrade, Jove Ilića, 154 Belgrade, Serbia
tamara.vlastelica@fon.bg.ac.rs

Abstract. Neuromarketing, a subfield of consumer neuroscience, investigates how consumers react to advertising stimuli using brain imaging and biometric techniques. It is a promising, multidisciplinary field facing practical and theoretical challenges, including limited theoretical frameworks, data complexity, and high research costs. Also, while the research contribution to the field of behavioral neuroscience is clear, the contribution to the field of marketing is far less so. A narrative literature review of 116 highly cited papers highlights both the opportunities and pitfalls of neuromarketing, offering insights into its current state and future prospects.

Keywords: neuromarketing · consumer behavior · consumer neuroscience

1 Introduction

The recognition of neuroscience's potential to enhance the study of consumer behavior dates back to 1979 (Kroeber-Riel, 1979). However, it was the development of noninvasive **neuroimaging** techniques, such as EEG and fMRI, that allowed the integration of neuroscience and consumer behavior research. Alongside these technological advancements, albeit at a slower pace, theoretical frameworks connecting brain activity to behavior have progressed. These frameworks now serve as the critical bridge between the theory and practice of consumer behavior, supported by empirical evidence generated through neuroimaging tools. The term "**neuromarketing**" first appeared in academic literature in 2007 (Lee et al., 2007; Fugate, 2007). However, the pioneering use of **neuroscience** in marketing is credited to Gerald Zaltman (2003), Professor Emeritus at Harvard Business School, who was the first to employ fMRI technology to study how brain activity responds to marketing stimuli. As advancements in brain imaging and biometric tools made these methods more accessible and affordable, and as computational capacity expanded to process vast datasets, neuromarketing evolved into a burgeoning field within both academia and industry.

In this paper, we explore the complexities, challenges, and concerns surrounding the use of neuromarketing techniques from both practical and theoretical standpoints.

V. Štavljanin et al. (Eds.): SymOrg 2024, LNNS 1681, pp. 40–49, 2025.
https://doi.org/10.1007/978-3-032-08090-5_4

We begin by outlining the primary methods and their defining characteristics. Following this, we examine the theoretical and methodological limitations, as well as the operational challenges that arise in practice. Our analysis of 116 highly-cited papers published between 2003 and 2023 investigates the consumer behavior theories most frequently studied, the neuroscientific methods employed, the brain regions involved, and the extent to which overarching theoretical frameworks are utilized to connect these findings. Finally, we differentiate neuromarketing from consumer neuroscience, highlighting scenarios where neuroimaging and biometric techniques are most applicable. Ethical concerns, though highly relevant, are not discussed here due to space constraints but remain critical and deserving of thorough examination.

2 Methods of Consumer Neuroscience and their Main Characteristics

Since the Roman physician Galen first identified the brain as the seat of behavior, there has been an enduring curiosity to observe its inner workings. However, it wasn't until 1924, when Hans Berger successfully recorded the brain's electrical activity, that this became possible without harming the subject (Panteliadis, 2021). Since then, numerous methods to study brain function have emerged, both direct and indirect. For this paper, we focus on two broad categories outlined by Cerf (2017): a) **brain imaging** tools that measure **brain activity** and **biometrics**, and b) other techniques that capture secondary, indirect signals like eye movements or skin responses.

Unlike Cerf's (2017) comprehensive list, we excluded several techniques from our analysis due to reasons such as invasiveness, being a variation of existing methods, or limited commercial availability. For instance, we removed Positron Emission Tomography (PET) and Electrophysiology for their invasive nature, Solid-State Topography (SST) and Magnetoencephalography (MEG) due to availability, and Transcranial Magnetic Stimulation (TMS) because it induces brain activity rather than measures it. The final set of tools includes functional magnetic resonance imaging (fMRI) and electroencephalography (EEG) for brain imaging, alongside Galvanic Skin Response (GSR), Eye Tracking (ET), the Facial Action Coding System (FACS), and the Implicit Association Test (IAT). Each method is described, scored on cost (ranging from below $1,000 for "low" to above $100,000 for "high"), and assessed for complexity based on whether programming or neuroscientific expertise is required.

Functional Magnetic Resonance Imaging (fMRI). fMRI detects changes in blood flow and oxygenation levels in specific brain areas during various cognitive tasks and behaviors. It is by far the most expensive methodology available, as the costs of a study can be hundreds of thousands of dollars. It also scores the highest on complexity. Yet, it is the method of choice for most research in the field for reasons that need to be clarified (more about it in the next section).

Electroencephalography (EEG). EEG measures electrical activity in the brain using electrodes placed on the scalp. It relies on the electrode's ability to detect changes in the electromagnetic field, and it was a much older technique and neuroscientific method of choice until fMRI came onto the scene. The costs of a study can be in the tens of thousands of dollars, making the technique affordable for commercial use. However, while it is easier to operate than an fMRI machine, the data treatment is still complicated enough to make the tool highly complex.

Eye Tracking (ET). ET measures and analyzes the movement and focus of a person's gaze. It is ideal for researching visual attention, various cognitive processes, and decision-making. The method became popular with the onset of the Internet and the need to understand the consumer's visual journey when interacting with a webpage. It ranks medium in terms of costs and complexities, and the costs are likely to drop further.

Galvanic Skin Response (GSR). GSR, also known as electrodermal activity, measures the electrical conductance of the skin, which changes due to sweat gland activity. The real-time tracking doesn't interfere with a participant's experience, making it relatively easy to set up and execute. GSR works well with other methods like eye tracking or EEG, providing a fuller picture of the emotional and cognitive response. It is a tried and true method that ranks low on costs and complexity.

Facial Action Coding System (FACS). FACS analyzes facial expressions based on the movement of individual facial muscles, and it is purportedly suitable for measuring emotional states and nonverbal communication. However, as the new paradigm of a constructed emotion (Barrett, 2017a; 2017b) supplants classical theories of emotions, the reliability of facial expressions to describe, let alone explain emotions, is being rapidly abandoned by researchers (Clark et al., 2020).

Implicit Association Test (IAT). IAT is used to assess the strength of associations between concepts, and it is a way of researching implicit biases such as prejudice, discrimination, and stereotypes. It works by presenting participants with pairs of words or images and asking them to categorize them as quickly as possible. By measuring the time it takes participants to make these associations, the IAT can reveal implicit biases that individuals may not be consciously aware of. IAT is the least expensive and the least complex tool in the neuroscientific toolbox. However, some critics claim that the method has reached the status of a fad and is used to make strong claims with weak evidence (Blanton et al., 2009) (Table 1).

3 Methodology and Analysis

In accordance with PRISMA guidelines (Page et al., 2021), we conducted a comprehensive systematic review of 116 peer-reviewed articles. The initial database search yielded 1,847 potentially relevant articles from Google Scholar, published between 2003 and 2023. Following the PRISMA flow diagram protocol, duplicate entries (n = 426) were first removed, leaving 1,421 articles for preliminary screening. Secondary screening excluded 1,305 articles due to insufficient data, irrelevant outcomes, methodological limitations, and non-English language. The final sample of 116 articles met all inclusion criteria (Table 2).

The objective was to answer the following: what consumer behavior theories are most commonly studied, what neuroscientific methods were deployed, and whether an overarching theoretical work was mentioned to tie it all together. We used Google Scholar

Table 1. Methods of consumer neuroscience

Methods	Description	Costs	Complexity	Applications
Functional Magnetic Resonance Imaging (fMRI)	Detects changes in blood flow and oxygenation levels in specific areas of the brain	High	High	Brain function during various cognitive tasks and behaviors
Electroencephalography (EEG)	Measures electrical activity in the brain using electrodes placed on the scalp	Med.	Medium	Brain function and activity patterns
Eye Tracking (ET)	Measures and analyzes the movement and focus of a person's gaze	Med.	Medium	Visual attention, cognitive processes, and decision-making
Galvanic Skin Response GSR	Measures changes in the skin's electrical conductance in response to emotional arousal or psychological stress	Low	Medium	Emotional arousal or psychological stress
Facial Action Coding System (FACS)	Analyzes facial expressions based on the movement of individual facial muscles	Low	Medium	Emotional states and nonverbal communication
Implicit Association Test (IAT)	Assesses the strength of associations between concepts	Low	Low	Implicit biases or attitudes

and applied the following two sets of keywords: "neuromarketing" and "neuroscience and consumer behavior." The two sets of keywords covered most of the research grounds we were interested in without being too broad. The search yielded 52 papers for "neuromarketing" and 64 for "neuroscience and consumer behavior." All but 9 papers from the "neuromarketing" batch and 18 papers out of the "neuroscience and consumer behavior" were literature review papers and hence were not analyzed. The following analysis was done on a total of 55 papers.

Table 2. Overview of key articles reviewed in the study.

	Neuromarketing		Neuroscience & Consumer Behavior		Total
	Count	%	Count	%	
Primary Research	3	5.77%	40	62.50%	43
Literature Review	43	82.69%	18	28.13%	61
Theoretical	6	11.54%	6	9.38%	12
Total	52		64		116

Consumer Behavior. A scientific discipline that studies how individuals, groups, and organizations select, purchase, use, and dispose of goods, services, experiences, or ideas (Solomon, 2012) - is vast in its field of inquiry and theoretical treatment. It draws on psychology, sociology, economics, anthropology, and neuroscience for inspiration in both methods and topics of investigation. As such, the research that falls under its umbrella is very diverse. This is why it was a surprise to see purchase intent and especially the willingness to pay (WTP) dominate the research we analyzed - 20 papers in total - were on such a subject. One explanation to why most research in **consumer neuroscience** focuses on willingness to pay (WTP) is because it directly relates to consumer decision-making and market dynamics, making it a critical factor for the practical side of marketing. Research in this area often draws on established economic theories, such as prospect theory, which examines how people evaluate potential losses and gains. Neuroscientific findings complement these theories by providing empirical evidence about the brain's response to pricing, enhancing the theoretical understanding of consumer behavior.

What was most striking about the cohort of papers we analyzed is that fMRI was the neuroscientific method of choice in as many as 25 out of 43 experiments. The reasons for this are unclear. Sure, it can provide detailed insights into brain activity, including areas of the brain deep inside the cavity. The non-invasive nature of fMRI and its relatively high spatial resolution makes it ideal for capturing complex brain responses in real-time. However, the high costs and complicated data treatment are making the method inaccessible to most researchers and universities

It is worth noting the noticeable absence of the overarching theoretical frameworks. This aligns with the "theory poor" side of the equation mentioned in the previous section. However, there was one notable exception: neuroeconomics. **Neuroeconomics** (Glimcher, 2003; Glimcher, 2009) seeks to understand how the brain processes information and makes decisions, particularly in economic contexts, as was mentioned in a third of the papers (21 out of 64). The connection between neuroeconomics and the field of consumer behavior is obvious: both fields share a common interest in understanding how individuals make decisions. Neuroeconomics, unlike most consumer behavior investigations, takes a more biological approach by examining the neural mechanisms underlying these decisions. However, what makes neuroeconomics so well adapted for the field of consumer neuroscience is the fact that as much as it applies neuroscience to economic behavior, it also applies economics to neuronal activity (Montague, 2007).

By combining these perspectives, neuroeconomics lends itself well as a "bridge" theory between the brain and the mind, at least the part that generates behavior, and specifically, the decision-making behavior.

4 Challenges with Neuromarketing Methods

Neuromarketing research faces three significant challenges: the need for more robust theoretical foundations (Voytek, 2022), the complex handling of large datasets generated by neuroscientific methods (Cerf, 2017), and the high costs associated with conducting these studies (Cardoso, 2022). These hurdles make it difficult to advance the field and integrate neuromarketing findings into practical applications effectively.

Neuroscience is frequently described as "data-rich and theory-poor" (Favela, 2021), a reflection of the immense complexity of the brain, often called the most intricate object in the known universe (Ackerman, S. (1992). Discovering the brain. Stevens, 1999). While numerous theories have emerged to explain brain function, such as the Bayesian brain theory, coordination dynamics, the free-energy principle, network theory, and neural reuse theory (Favela, 2021), no single framework has gained universal acceptance. This lack of theoretical consensus is particularly challenging for researchers, especially those outside neuroscience, like many in neuromarketing. Lee (2018) and colleagues have identified gaps in methodological understanding, over-reliance on reverse inference (Poldrack, 2006; Plassmann et al., 2012), and assumptions about functional localization of behavior, despite insufficient evidence (Friston, 2000).

The "data-rich" nature of neuromarketing studies presents significant challenges for researchers. For instance, a typical fMRI **experiment** with just 20 participants can generate several gigabytes of time-series data, even after removing noise, normalizing for different head sizes, and filtering out irrelevant signals (Cerf, 2017). Even though both marketing researchers and practitioners are accustomed to dealing with big datasets (Zekavica et al., 2018; Vukmirović et al., 2018), managing and interpreting data coming out of fMRI scans often requires expertise and resources that many marketing scholars may lack. Moreover, this data complexity ties into the broader "replication crisis" across scientific fields, where researchers struggle to replicate the empirical results of highly cited studies. Since the early 2000s, concerns have been raised about the reliability of many peer-reviewed studies, as researchers from disciplines such as psychology and genetics have frequently failed to replicate influential findings (Ioannidis, 2005; Baker, 2016; Open Science Collaboration, 2015; Moonesinghe, 2007; Altmejd, 2019). While the replication crisis affects various scientific domains, its implications should prompt neuromarketing researchers, especially in academia, to critically assess the robustness of their methodologies.

Neuromarketing techniques come with significant costs. In the U.S., a standard fMRI study can cost around $100,000, with expenses largely driven by equipment rental and technical staffing. EEG studies, though more affordable, can still reach up to $20,000. These high costs may explain why most neuromarketing research is concentrated in Europe and China (Cardoso, 2022). Advances in consumer tech, such as Apple's Vision Pro, which integrates eye-tracking and biometric sensors, are expected to reduce costs and broaden accessibility. However, despite cheaper alternatives like biometric tools, these methods still can't rival traditional, cost-effective surveys in terms of scalability. Additionally, high study costs often result in low sample sizes, sometimes fewer than 20 participants, which diminishes statistical power and hinders the detection of meaningful effects in scientific research (Button, 2013).

5 Opportunities for Neuromarketing

The challenges mentioned are not a condemnation of neuromarketing but rather a reminder to approach it cautiously as it gains traction in academic and commercial spheres (Plassmann et al., 2015a, 2015b; Robaina-Calderín, 2021). In the remainder of this paper, we explore scenarios where neuromarketing methods are not just useful but essential, such as when studying real-time responses to stimuli, subconscious reactions, or validating findings from conventional research techniques.

Traditional marketing research - whether quantitative (surveys) or qualitative (interviews) - relies heavily on subjects recalling and reporting their experiences after the fact, which introduces a critical flaw (Macdonald, 2012). The disconnect between how people feel in the moment and what they report later is particularly relevant in assessing advertising effectiveness (Jovanovic et al., 2016; Ohme et al., 2009; Stipp & Woodard, 2011). Neuromarketing offers a way to mitigate this flaw by capturing more immediate, authentic reactions.

Traditional marketing research often depends on subjects to self-report their thoughts and feelings about a stimulus, utilizing methods like interviews, focus groups, and surveys (Shukla, 2008; Fugate, 2007). This reliance on self-reporting aligns with the prevailing view of individuals as rational agents, encapsulated in the concept of "homo economicus" (Singer, 2013). Also, there has been a highly prevalent school of thought that individual intentions and behavior are conditioned by previously formed attitudes (which have their cognitive, affective, and conative components) based on experience with or exposure to an object or stimulus (Vlastelica et al., 2018). However, research by behavioral economists such as Daniel Kahneman demonstrates that human decision-making is often less rational and more biased, highlighting the significance of implicit attitudes and subconscious processes (Kahneman & Tversky, 1982; Kahneman, 2011; Plassmann, 2015; Vlastelica et al., 2023). Neuroimaging and biometric techniques are particularly effective in revealing these underlying processes. For example, they can shed light on the impact of emotions on decision-making (Pavkovic et al., 2023; Damasio, 1996), the role of cognitive dissonance in belief formation (Harmon-Jones, 2019), and the challenges individuals face in articulating their emotions (Russell, 2009; Barrett, 2017). Understanding these phenomena is crucial in marketing (Vlastelica, 2016), yet traditional self-reporting methods fail to capture them effectively. Additionally, there exists a substantial body of theoretical and empirical research that facilitates the mapping of behaviors to specific brain regions or biometric signals, offering deeper insights into consumer behavior.

Finally, neuromarketing methods offer a crucial complement to traditional marketing approaches by validating their results and enhancing the reliability of findings. As Plassman (2015) and his team point out, consumer neuroscience should be seen not as a replacement for classic behavioral measurements but as a valuable addition that refines and enhances the accuracy of interpreting those measures. Venkatraman et al. (2015) further demonstrated this by comparing traditional self-reports with neuromarketing methods like eye tracking, biometrics, EEG, and fMRI in predicting advertising effectiveness, showing how neuromarketing can strengthen the interpretation of real-world marketing success. Additionally, these techniques can contribute to better theory-building efforts, as highlighted by Alvino (2019), thus reinforcing the importance of neuroscience in both practical applications and academic insights.

6 Conclusion

This paper discusses three primary challenges facing neuromarketing research: the need for more robust theoretical frameworks, the complexity of managing large datasets, and the high costs associated with these studies. Neuroscience has often been criticized for its lack of unified theories about how the brain functions, which poses difficulties for neuromarketing, as it requires a clear connection between consumer behavior and brain activity. Additionally, the vast data generated in neuromarketing studies requires significant resources to process, and the expenses involved in conducting such research remain a major hurdle. However, neuromarketing can offer unique insights, particularly in cases where real-time feedback, the exploration of subconscious attitudes, or validation of existing consumer behavior theories is necessary. There is also reason for optimism as advancements in neuroscience continue and the costs of neuromarketing technology decrease, allowing for more widespread use. Despite the obstacles, neuromarketing is still in its early stages, with considerable potential for growth and development.

Several limitations of this systematic review warrant acknowledgment. The primary constraint stems from our exclusive reliance on Google Scholar as the sole database for literature retrieval. Additionally, our search strategy employed a relatively narrow set of predefined keywords without exploring semantic variations, synonyms, or related terminologies, potentially excluding pertinent studies that used alternative terminology or conceptual frameworks. Feature research should include Web of Science and other resources, and a combination of keywords should be tested to ensure all relevant studies are considered.

References

1. Altmejd, A., et al.: Predicting the replicability of social science lab experiments. PLoS One. **14**(12), e0225826 (2019)
2. Barrett, L.F.: How emotions are made: The secret life of the brain. Houghton Mifflin Harcourt (2017)
3. Barrett, L.F.: The theory of constructed emotion: an active inference account of interoception and categorization. Soc. Cogn. Affect. Neurosci. **12**(1), 1–23 (2017)
4. Blanton, H., Jaccard, J., Klick, J., Mellers, B., Mitchell, G., Tetlock, P.E.: Strong claims and weak evidence: reassessing the predictive validity of the IAT. J. Appl. Psychol. **94**(3), 567 (2009)
5. Button, K.S., et al.: Power failure: why small sample size undermines the reliability of neuroscience. Nat. Rev. Neurosci. **14**(5), 365–376 (2013)

6. Cardoso, L., Chen, M.M., Araújo, A., de Almeida, G.G.F., Dias, F., Moutinho, L.: Accessing neuromarketing scientific performance: Research gaps and emerging topics. Behavioral Sciences. **12**(2), 55 (2022)
7. Cerf, M.: Consumer neuroscience. Mit Press (2017)
8. Clark, E.A., et al.: The facial action coding system for characterization of human affective response to consumer product-based stimuli: a systematic review. Front. Psychol. **11**, 920 (2020)
9. Damasio, A.R.: The somatic marker hypothesis and the possible functions of the prefrontal cortex. Philos. Trans. R. Soc. Lond. B Biol. Sci. **351**(1346), 1413–1420 (1996)
10. Favela, L.H.: Fundamental theories in neuroscience: why neural darwinism encompasses neural reuse, pp. 143–162. New Challenges in the Philosophy of Neuroscience, Neural Mechanisms (2021)
11. Friston, K.J.: The labile brain. I. Neuronal transients and nonlinear coupling. Philos. Trans. R. Soc. Lond. B Biol. Sci. **355**(1394), 215–236 (2000)
12. Fugate, D.L.: Neuromarketing: a layman's look at neuroscience and its potential application to marketing practice. J. Consum. Mark. **24**(7), 385–394 (2007)
13. Glimcher, P.W.: Decisions, uncertainty, and the brain: The science of neuroeconomics. MIT Press (2003)
14. Glimcher, P.W., Camerer, C.F., Fehr, E., Poldrack, R. A. (eds.).: Neuroeconomics: Decision Making and the Brain. Elsevier Academic Press (2009)
15. Harmon-Jones, E., & Mills, J. (2019). An introduction to cognitive dissonance theory and an overview of current perspectives on the theory.
16. Ioannidis, J.P.: Why most published research findings are false. PLoS Med. **2**(8), e124 (2005)
17. Jovanovic, P., Vlastelica, T., Cicvaric Kostic, S.: Impact of Advertising Appeals on Purchase Intention. J. Manag. Theory Pract. **21**(81), 35–45 (2016)
18. Kahneman, D.: Thinking, fast and slow. Macmillan (2011)
19. Kahneman, D., Tversky, A.: The psychology of preferences. Scientific american. **246**(1), 160–173 (1982)
20. Kroeber-Riel, W.: Activation research: Psychobiological approaches in consumer research. J. Consum. Res. **5**(4), 240–250 (1979)
21. Lee, N., Broderick, A.J., Chamberlain, L.: What is 'neuromarketing'? A discussion and agenda for future research. Int. J. Psychophysiol. **63**(2), 199–204 (2007)
22. Lee, N., Chamberlain, L., Brandes, L.: Welcome to the jungle! The neuromarketing literature through the eyes of a newcomer. Eur. J. Mark. **52**(1/2), 4–38 (2018)
23. Macdonald, E., Wilson, H.N., Konus, U.: Better customer insight-in real time, vol. 90. Harvard Business School Publishing (2012)
24. Montague, P.R.: Neuroeconomics: a view from neuroscience. Funct. Neurol. **22**(4), 219 (2007)
25. Moonesinghe, R., Khoury, M.J., Janssens, A.C.J.W.: Most published research findings are false—but a little replication goes a long way. PLoS Med. **4**(2), e28 (2007)
26. Ohme, R., Reykowska, D., Wiener, D., Choromanska, A.: Analysis of neurophysiological reactions to advertising stimuli by means of EEG and galvanic skin response measures. J. Neurosci. Psychol. Econ. **2**(1), 21 (2009)
27. Open Science Collaboration: Estimating the reproducibility of psychological science. Science. **349**(6251), aac4716 (2015)
28. Page, M.J., et al.: The PRISMA 2020 statement: an updated guideline for reporting systematic reviews. BMJ. **372**, 71 (2021)
29. Panteliadis, C.P.: Historical overview of electroencephalography: From antiquity to the beginning of the 21st century. J. Brain Neurol. Disord. **3**, 1–10 (2021)
30. Plassmann, H., Ramsøy, T.Z., Milosavljevic, M.: Branding the brain: A critical review and outlook. J. Consum. Psychol. **22**(1), 18–36 (2012)

31. Plassmann, H., Venkatraman, V., Huettel, S., Yoon, C.: Consumer neuroscience: applications, challenges, and possible solutions. J. Mark. Res. **52**(4), 427–435 (2015)
32. Plassmann, H., Venkatraman, V., Huettel, S., Yoon, C.: Consumer neuroscience: applications, challenges, and possible solutions. J. Mark. Res. **52**(4), 427–435 (2015)
33. Poldrack, R.A.: Can cognitive processes be inferred from neuroimaging data? Trends Cogn. Sci. **10**(2), 59–63 (2006)
34. Robaina-Calderín, L., Martín-Santana, J.D.: A review of research on neuromarketing using content analysis: key approaches and new avenues. Cogn. Neurodyn. **15**(6), 923–938 (2021)
35. Russell, J.A.: Emotion, core affect, and psychological construction. Cognit. Emot. **23**(7), 1259–1283 (2009)
36. Shukla, P.: Essentials of Marketing Research. Bookboon, Denmark (2008)
37. Solomon, M., Russell-Bennett, R., Previte, J.: Consumer behaviour. Pearson Higher Education AU (2012)
38. Stevens, C.F.: Complexity of brain circuits. In: Complexity: Metaphors, Models, and Reality, pp. 245–261 (1999)
39. Vlastelica, T.: Korporativna reputacija i društvena odgovornost. Zadužbina Andrejević (2016)
40. Vlastelica, T., Cicvarić, K.S., Okanović, M., Milosavljević, M.: How Corporate Social Responsibility Affects Corporate Reputation? The evidence from an emerging market, JEEMS-Journal of East European Management Studies. **23**(1), 6–25 (2018)
41. Vlastelica, T., Kostić-Stanković, M., Krstić, J., Rajić, T.: Generation Z's intentions towards sustainable clothing disposal: Extending the theory of planned behavior. Pol. J. Environ. Stud. **32**(3), 2345–2360 (2023)
42. Voytek, B.: The data science future of neuroscience theory. Nat. Methods. **19**(11), 1349–1350 (2022)
43. Vukmirović, A., Zoltan, R., Radojičić, M., Vukmirović, J., Jovanović Milenković, M.: Infrastructural Model for the Healthcare System based on Emerging Technologies. Acta Polytechnica Hungarica, Journal of Applied Studies. **15**(2), 33–48 (2018)
44. Zaltman, G.: How customers think: Essential insights into the mind of the market. Harvard Business Press (2003)
45. Zekavica, A., Djordjevic, A., Vukmirovic, A., Vukmirovic, J., Brankovic, M.: The Potential of The Big Data Implementation in Serbia in the context of Marketing Research. Management: Journal of Sustainable Business and Management Solutions in Emerging Economies. **23**(2), 39–48 (2018)
46. Alvino, L.: How can we improve consumer behaviour research? A critical literature review on the contributions and the limitations of consumer neuroscience. Proceedings of the 33rd International Business Information Management Association Conference, IBIMA, 5947–5951 (2019)
47. Baker, M.: 1,500 scientists lift the lid on reproducibility. Nature. **533**(7604) (2016)
48. Pavković, V., Vlastelica, T., Kostić-Stanković, M., Dlačić, J.: Experiences and expectations: drivers of urban tourist destination competitiveness. Manag.: J. Sustain. Bus. Manag. Solut. Emerg. Econ. (2023). https://doi.org/10.7595/management.fon.2023.0016
49. Singer, T.: Beyond the Homos Economicus. The Project Sindicate https://www. project-syndicate. org/commentary/a-new-model-of-human-behavior-bytania-singer (2013)
50. Stipp, H., Woodard, R.P.: Uncovering emotion: Using neuromarketing to increase ad effectiveness. Advertising Research Foundation. **9** (2011)

Harnessing Social Networks for Brand Strategy

Damir Vasilj and Željko Matković(✉)

University of Mostar, Trg hrvatskih velikana 1, Mostar, Bosnia and Herzegovina
zeljkomatkovic@hotmail.com

Abstract. This paper synthesizes the strategic shift that occurred in branding in the digital era, wherein it transitioned from traditional methods of branding toward one shaped by digital platforms and social media. It emphasizes how Twitter, Instagram, and other digital venues have broadened the definition of consumer-brand touch points, and how, therefore, engagement is no longer a good thing to do but an imperative. Brands have shifted from one-way communication to real-time, interactive exchanges that cultivate authentic ties with their audience. Integration of user-generated content and social media influencers in the branding efforts helped brands establish credibility with, and engage new groups. This paper emanates from challenges that, in this fast-changing digital landscape, brands are facing: Unstable social media algorithms, privacy issues, and to a large extent, material oversaturation. The organizations have to adapt and follow methodologies such as SEO, content marketing, personalized advertising, among others, in order to keep the audience tuned. Up-and-coming trends that will no longer be merely personalization driven by AI, visual search, and augmented reality, but a change in the way consumers will interact with the brands. All of these will maintain a sensitive balance with core values, privacy concerns, and shifting consumer expectations that the brands will be expected to balance. It also pointed out the importance of adopting agile methodology, gave priority to consumer-oriented branding strategies, and innovation sustained in the ever-evolving digital market. This would mean that even more research should be carried out in order to fully understand the long-term effects of new technology on branding.

Keywords: Brand Engagement · Social Media Influencers · User-Generated Content · Digital Branding · Personalization

1 Introduction

The current condition of branding illustrates how the digital era can transform possibilities. In the last fifteen years it is witnessed extensive utilization and accessibility of social media applications, resulting in a novel domain of **interaction** between customers and companies, which is characterized by highly engaged and interconnected young. In view of the benefits and challenges this changing environment offered for brands, conventional branding techniques would need to be reevaluated.

This paper provides a broad overview of current **digital branding** strategies, which are going through a shift from traditional methods to methods that are shaped by social

V. Štavljanin et al. (Eds.): SymOrg 2024, LNNS 1681, pp. 50–61, 2025.
https://doi.org/10.1007/978-3-032-08090-5_5

media and digital platforms. The paper assembles findings from recent science papers, while taking a panoramic view of how brands can navigate through digital **interaction**, **user-generated content**, and digital influencers.

Ever since digital marketing first appeared in the internet age, brands have come to understand how crucial it is to project a professional image. This included the exploitation of brand methods during this period (Kingsnorth, 2022). With increasingly more audiences turning online for entertainment and to acquire information, brands are compelled to evolve their communication approach in an attempt to connect with their target audiences. Social media advertising, content marketing and **search engine optimization** SEO have now become important catalysts of brand communication (Schwartz, 2021)

One of the most significant shifts in brand positioning tactics has likely been the move from one-way communication to interactive involvement. Unlike traditional media, which allows for general communication, social media platforms allow for real-time communication between the brand and its customers. This strengthens and improves the relationship between the brand and its customers (Verhoef, et al., 2010). As involvement becomes the norm brands approach communication in a much more humane way, forming genuine connections with customers as opposed to merely disseminating information (Bijmolt, et al., 2010).

Finally, **user-generated content** and **social media influencers** bring the human side into modern social media. The consumers are no longer passive receivers of content that has been produced in the name of business; instead, they become active coproducers and co-sharers of the content about the brands that they choose (van Doorn, et al. 2010). Additionally, brands took advantage of the effect and influence of social media to reach new audiences with their message. Influencers collaborate with individuals who have substantial and engaged fan bases (Kim, and Johnson, 2016).

Nevertheless, these opportunities come with a corresponding set of challenges. In the light of opportunities and threats such a changing environment presented to brands, traditional practices of branding would be thoroughly questioned. Such examples include content overload, loss of privacy, and the forever-changing algorithms of the social media platform (Puzakova, and Aggarwal, 2018). As discussed by Lang and Kenny it is only then that the brand would remain highly relevant in the market, considering the very high speed with which technological advancement is being experienced, and the brand has to be highly adaptive and flexible to the rate at which new advancements and trends in the industry would be embraced (Lang, and Kenny, 2024).

This paper will focus on how branding approaches have evolved in the social network era in response to its development in last fifteen years. The paper will take a careful look at how companies use social media to communicate and engage with their clientele, as well as what happens when fresh problems and fashions emerge. Ultimately, by being more knowledgeable about contemporary branding strategies, a firm would position itself to compete in the rapidly expanding digital space.

The methodology undertaken for this paper is a conceptual analysis supported by a review of the literature on branding in light of the digital era, **social media engagement**, and consumer-brand **interaction**. The authors of this paper rely on secondary sources in synthesizing understandings of various researches on how digital platforms, social networks, and emerging technologies affect the strategizing of branding. This paper examines theoretical frameworks to analyze how brands can engage consumers and how they build loyalty through social media platforms.

The emphasis of the review lies in **user-generated content**, influencer marketing, and algorithm-driven content strategies. Through a careful review, the authors shed light on the best practices and challenges involved and position the study in the broader context of **digital branding** research.

Authors will try to answer on how brands can effectively use social networks to engage better with consumers, build brand loyalty and navigate through emerging challenges in the digital age.

The paper starts by exploring how digital transformation is reshaping branding, while marking a clear shift from traditional, more static methods to those which are fueled by digital platforms and social media. It traces the way branding strategies have changed over time, showing how brands have moved from a one-way approach (simply delivering messages) to creating interactive experiences where audiences can engage directly and meaningfully.

Then, the paper goes further in describing the disruption of brand communication by social networks. Later will be explained how **user-generated content** and social media influencers support improving brand awareness by building trust with consumers.

It then delves into how to extend consumer engagement by stimulating much more audience activity. "Brands can enhance their connectivity with customers by making content interactive or through gamification features, which is the rise of users' involvement in online activities.

It also looks through trends and challenges which brands are facing in the digital age. These trends are including the adoption of advanced technologies like **artificial intelligence**, **augmented** and virtual **reality** and tailored and personalized content. Challenges of managing through content overload, challenge of staying authentic, and respecting privacy are also addressed in this paper.

Finally, the paper is reflecting on the ongoing evolution of brand-consumer relationships, while emphasizing that to thrive in today's fast-moving digital landscape, brands need to be adaptable, innovative, and genuinely focused on their customers.

2 Evolution of Branding Strategies

In view of the benefits and challenges this changing environment offered for brands, conventional branding techniques would need to be reevaluated. This shifting environment provided brands both benefits and problems, thus conventional branding approaches must be reevaluated.

Branding has shifted from its conventional approaches to the dynamic strategies pushed by social networks, as this article will demonstrate. As digital marketing began, it became evident that the internet's rapid expansion made it difficult for companies to get recognized (Kingsnorth, 2022). Eventually, with the growth of the internet, brands soon realized that they needed to be online. Thereafter, digital media became a prevailing mode of communication for brands to reach out and affect their target groups (Schwartz, 2021).

Branding methods changed with social media. Social media platforms enable dynamic **interaction,** changing brand-customer relationships (Verhoef, et al., 2010). Companies had to switch from one-way broadcasting to interactive communication. Brand live communication conversations entered a new age with such consumer **interaction.**

Contemporary branding tactics emphasize commitment and building lasting relationships with consumers through several touchpoints. Social media lets brands talk to customers about community building and relationships. By emphasizing **interaction**, transactional communications seem to have been replaced by relationship-building beyond corporate transactions.

In the social media, the introduction of **social media influencers** and **user-generated content** marked the revolution in marketing communication. Consumers co-create, develop, and share relevant content of the brand (van Doorn, et al. 2010). This makes UGC (**user-generated content**) an asset for brands as it allows them to reach more people and build a real relationship with their customers, blurring the lines between the brand and the consumer.

Through **user-generated content**, **social media influencers** have become a brand exposure force. Powerful people affect others and their followers. Company strategic partnerships use these powers to boost visibility and message credibility (Kim, and Johnson, 2016). Marketing with influencers gives companies access to targeted audiences and personal connections with customers. Branding hasn't improved completely. Content-heavy ecosystems with privacy issues and algorithmic uncertainty require companies to negotiate (Puzakova, and Aggarwal, 2018).

Companies must adapt to technology to generate new ideas and trends that can be accepted later (Lang, and Kenny, 2024).

Brand strategy is requiring flexibility and strength. The rapid digital marketing growth and social media age are call for updating the language of communication in order for brands to keep up with consumer preferences in fashion and technology.

3 Leveraging Social Networks for Brand Communication

Social media has transformed and still is transforming brand communication. More than ever before, brands can connect with their audience and potential consumers in dynamic ways. Brands can utilize social media to develop relationships, to engage potential consumers, and to promote their message.

Social media allows brands and customers to interact in real time. Interactive social networks allow brands to develop communities, launch conversations, and instantly respond to customer inquiries and comments (Verhoef, et al., 2010). Communication helps brands appear more personable and connect with their audience. This creates loyalty and trust.

Speaking of brand communication one term that has to be mentioned is "word of mouth". In addition to receiving information through word-of-mouth face-to-face **interactions** with others, consumers in the digital age can learn from product reviews on Internet opinion sites such as Amazon, Yelp, Trustpilot or social media like Facebook, Twitter (Bartschat, et al., 2022). In their study conducted by Oscarius and others in 2021.

using SmartPLS analysis was shown that Word-of-Mouth has no significant effect on purchase of Smartphones. Therefore, they concluded that by increasing Word-of-Mouth there is no significant increase on purchase intention. (Yudhi, et al., 2021). In their study Bartschat, M. and others find in their research that consumers typically turn to face-to-face communication and internet opinion websites early in the decision-making process to gather foundational information. (Bartschat, et al., 2022). This suggests that resource availability is a strong driver of which word of mouth type consumers use for product information. It is furthered by the viral aspect of brand communication, in which appealing content such as interesting videos, stories, or other testimonials is spread far and wide across the networks through sharing, likes, and comments by the users.

Social media gives marketers many content types and plot tools to communicate clearly. Marketers exploit social media's multimedia features to engage their audience (Bijmolt, et al., 2010). This contains stunning photographs, engrossing films, and captivating stories. Making content that people want to share helps brands reach more people across digital channels.

Along with free content, corporations utilize social media advertising to attract specialized audiences. Facebook, Instagram, and Twitter allow corporations to target specific groups based on their hobbies, behaviors, and demographics with strong advertising tools (Kingsnorth, 2022). Targeted advertising efforts can boost brand awareness and digital traffic. This helps them achieve their goals and maximize ROI.

Brands may now use **social media influencers** to boost their online presence. People with many fans are influencers. Their trustworthiness and authenticity help brands convey brand messages and engage with consumers (Kim, and Johnson, 2016). Influencer relationships help brands reach niche audiences and build personal **interactions** with customers. This boosts brand loyalty and interest.

Social media lets brands leverage **user-generated content** to promote their business. Brand stories are enhanced with consumer content regarding their brand experiences, preferences, and connections (van Doorn, et al. 2010). **User-generated content** helps brands reach more people, form relationships, and create a community. This blurs brand-consumer boundaries.

Brands face issues like too much content, unexpected algorithms, and shifting consumer preferences (Puzakova, and Aggarwal, 2018). Social media platforms are continually changing, therefore brands must be flexible and ready to adapt to new trends and technology (Lang, and Kenny, 2024).

Brands can assist their potential consumers in communicating with each other by creating online communities or social media groups and discussion forums around their products. This way brands can encourage **interactions**, they can ask questions, host live events, or even respond to comments. Brands can share **user-generated content** through testimonials or customer stories that could initiate conversations while helping potential consumers to get a peek at how products are really applied. It is through the provision of such platforms that one is able to engage in a collaborative space wherein consumers share insights, go forward with questions, and make recommendations for the purpose of cementing brand loyalty and communities of like-minded individuals.

Brand promotion on social media can be essential for companies which are digitally successful. By harnessing social media's interactive features while providing engaging content and if they are partnering with influencers, brands can engage with potential and current consumers and promote their message throughout digital channels. Brands must continue to deliver value, to encourage connections, and thrive to maintain long-term relationships with their audience as they improve and develop new social media strategies.

4 Enhancing Consumer Engagement

Consumer engagement is the foundation of modern branding strategies and represents the dynamic relationship that exists between brands and their target audience. Here will be examined the various approaches that companies and brands can employ to enhance consumer **interaction** in the digital environment while drawing on insights from research and journals.

Customer engagement through social media represents a lot more than passive consuming. Essentially, this is a description of how brands can have social media in an interactive way with their audience and how can they reach the market segment with intimate offers of eloquent **interaction** through real conversation and interactive engagement (Calder, et al., 2015). Brands which are actively engaging can create trust and sense of belonging for their potential and current consumers.

Whilst modern branding strategies are becoming more reliant on social media it can provide companies and their brands a direct connection to their consumers. In this way brands can better understand consumer behavior even in real time. Social media marketing has really enabled brands to create deeper emotional connections with consumers and build brand loyalty across a range of platforms.

Through social media marketing across various platforms, brands are able to build more long-lasting loyalty and deeper emotional bonds with their customers. By tapping into the relationships between users on social media, brands can boost their visibility and strengthen customer loyalty (Libai, et al., 2010). Companies can transform social networks into strategic assets by prioritizing these **interactions** and incorporating them into their broader marketing strategies (Libai, et al., 2010).

One of the key benefits of leveraging social media in branding involves its capability to analyze vast volumes of consumer data. Using the volume, variety, and speed of social media data, a firm could find out about customer preferences and behavior to create very targeted marketing messages. In the fashion industry, emotional engagement with a brand can trigger impulse purchases, which has also been demonstrated to influence consumer purchasing behavior through social media marketing. Consumers' online buying behavior is heavily influenced by social media activities that highlight brand resonance and emotional connection (Safeer, 2024). This underlines the relevance of emotional engagement in social media branding since it is linked not just to loyalty but also to immediate purchase.

The significance of the rise of interactive content formats is that they create new innovative ways of reaching out to their target audience and bring them closer to experiencing their brand. In order to improve experience of current and potential customer and to develop deeper relationships, brands can use a wide range of content forms, such as interactive polls, immersive movies, live streaming events, surveys, etc. (Baer,

2024). Through offering of entertaining and educational content marketers are trying to encourage audience **interaction** and customer feedback. This way it can boost consumer engagement and strengthens message of their brand.

With gamification, the most passive consumers can be turned into active players by a process of incrementally increasing engagement. Goyal and Verma in their study indicate that in the integrative setting, brand loyalty mediates the relationship between brand engagement and OBE (overall brand equity), and OBE mediates the relationship between brand loyalty and purchase intention (Goyal, and Verma, 2024).

Social media influencers bring validity with a power that brings forward the best in brand communication and involvement of consumers. Brands influence whose audience is similar to theirs to provoke conversations, excite, and build genuine relationships (Shao, 2023). Relationship influence thus provides brands not only the reach to a certain set of unique audiences but also in-person interactivity that basically supercharges consumer engagement and drives brand advocacy.

What is more important to understand is that **user-generated content** enhances customer engagement with the power to work as a very strong stimulus. The thing is that customers are allowed to build up the brand's story according to their experience, storyline, and, more importantly (Gambetti, et al., 2016). Opportunity to harness **user-generated content** for more mileage, real connections, and a community feel. This will see the brand acknowledging and featuring the **user-generated content** that empowers the customers and encourages them to become brand ambassadors, thus user engagement and loyalty.

However, for all of the opportunities and provocations customer **interactions** bear, marketers must tread slightly more challenging landscape, speckled with a host of obstacles, including oversaturation of material, skepticism about authenticity, and even doubt with algorithms (Lang, and Kenny, 2024). Besides, with changing consumer preferences, the brand has to be flexible and changeable to an extent where it can be changeable enough to modify its game plans in consonance with the rapidly changing needs and expectations of its customer base.

Strategic adoption of **artificial intelligence** enables businesses to outperform their competitors and achieve greater levels of success by producing creative solutions, providing a tailored customer experience, and opening up new revenue sources (Vasilj, et al., 2024).

Boosting consumer engagement is a critical strategy for brands aiming to thrive in today's digital age. Through interactive content such as gamification, using **social media influencers**, and presenting well defined **user-generated content**, brands are enhancing their communication on digital platforms. This way they can provoke deeper engagement and are able to build stronger and more meaningful connections with their potential consumers. In today's situation where consumer engagement is constantly changing it is a must for brands to stay committed on delivering value, building connections, and developing lasting relationships with their consumers.

5 Emerging Trends and Challenges

The changing dynamics of branding strategies greatly mold the trajectory of dynamism developing in **interactions** between brands and consumers within the dynamic field of **digital branding**. Thus, in this segment, an analysis into such changing dynamics of branding strategies will be provided through reviewing emergent trends, challenges to fixing, or problems. It does that through referring to scholarly research and industry practice. These papers or books have authors who consider advancement in trends. Some of the advancement in trends will be explained below.

Two features thar are currently revolutionizing the way they allow potential and current consumers to interact into more interactive user experiences are **augmented reality** (AR) and virtual reality (VR) (Baer, 2024). Virtual and **augmented reality** can be utilized by brands in order to develop more enhanced than conventional experience. Through new technologies brands can truly engage consumers with their products and services in completely new ways.

Fundamental aspects of modern branding strategies became customization and **personalization**. Through this approach companies provide customized experience and can connect deeply with every consumer individually. (Lang, and Kenny, 2024). Thanks to that, through the use of **artificial intelligence** algorithm and data-based insights, brands can customize content, recommendations, and offers to the taste and particular needs of every individual consumer. Marketing is one of the domains in **Artificial Intelligence** and is making changes to an exceptional extent (Vasilj, et al., 2023). **Artificial Intelligence** implementation in marketing has several advantages; the main one is that business organizations are able to deliver as high-level customer satisfaction with this technology by providing better services and adding value efficiently performing various marketing processes and managing data sensitivity.

Several authors raised the question of what makes an employee good, how does the organization go about finding such employees, and how does the organization engage in encouraging and participating in their development and in raising the quality of work (Vasilj, et al. 2022). In marketing, it has utilized AI to track down top performing employees by analyzing sales data, consumer **interaction**, and campaign results. These insights can thus enable the companies to focus on those factors that contribute much towards their success. Artificial intelligence will also help employees grow by providing them with personalized training in the sphere of digital marketing and customer engagement, which will ensure continuous improvement in customer satisfaction and efficiency of campaigns.

In their marketing strategies, brands prioritize diversity and inclusivity to reflect the cultural nuances and different demographics of their target audience (Gil, 2019). Inclusive marketing practices aim at creating a sense of belongingness among all types of consumers through the embracing of diversity and the challenging of stereotypes.

Brands are using **artificial intelligence** in order to allow organizations to predict search engine trends and analyze large amounts of data (Enge, et al., 2023). The most important reason behind such usage is that **artificial intelligence** tools are helping organizations understand the nature of the digital world in order to maximize their reach and increase the position of their existence on online search engines. Nowadays **artificial intelligence** is taking more and more space in marketing. One of the fields where it is being used the most is strategic planning. It can be said that **artificial intelligence** is not just some fancy tech word, it is definitely becoming one of the components needed for success in fast-moving digital world (Vasilj, et al. 2023).

Search engine optimization (SEO) is also changing nowadays. One of the main aspects of change is visual search, where consumers can find information on searches via images and not just on inquiries which are based on text (Enge, et al., 2023). To be sure that they can be more easily visible in visual search outcomes, digital assets such as videos and photos are being enhanced by organizations.

Due to effective management of Brand SERPs (Search Engine Results Pages), companies can enhance the visibility, credibility, and trustworthiness of their brands in the minds of online users who seek relevant information (Barnard, 2022). This is even though there are a lot of improvements in trends and one could think that the new trends are somehow an indomitable trend, but still, there remain lots of challenges if one takes into consideration that it is still a relatively unknown area regarding legislative, privacy, etc.

Information overload and **content saturation** is one of the challenges. The continuous growing amount of material on digital platforms has led to **content saturation**, which is making it more difficult for brands to distinguish themselves and engage consumers. (Puzakova, and Aggarwal, 2018). This is the reason why brands should develop high-authority content that has very high relevance for the target audience.

Authenticity and trust are some of the most difficult challenges a brand can face in today's digital context, where the consumer finds himself constantly massed by sponsored content and partnerships of all kinds (Gil, 2019). Transparency, honesty, and integrity should come first among the brand's ways of communicating to the consumer in an attempt to gain and build his trust.

Increasing knowledge about privacy and data security was revealing the need to follow ethical data practices and regulatory requirements, like the General Data Protection Regulation (GDPR) (Lang, and Kenny, 2024). Brands have an obligation to respect consumer privacy and protect consumer data. The failure to do so may undo risks and discredit from the target demographic on a brand. Challenges lie in the collection, storage, and use of consumer data for brands, considering the rigorous enforcement of data privacy regulations like the General Data Protection Regulation (GDPR) (Ramaswamy, and Ozcan, 2016). Brands should be in front of their transparent and ethically acting data within regulations, aiming only at further establishments and maintaining consumer trust.

Brands, however, have to manage the changing consumer preferences, which follow fast-changing market conditions and technological progressions effectively, if they have to remain in competitive edges (Ramaswamy, and Ozcan, 2016). With the rise of the digital age, brands had to adjust to changing needs and expectations from the consumer in order to remain relevant and engaging.

The biggest challenge that brands are facing in a really effective way to facilitate smooth omnichannel experiences is balancing the offline with online i**nteraction** points. (Hsieh, and Kumar, 2023). Brands will also have to find a subtlety threshold in touchpoints that would be specific to varying tastes and behaviors of consumers.

Brands will have to engage in a strategic, proactive approach in dealing with these emergent trends and challenges in **digital branding**. Other key success factors of the brand name to compete in this ever-chaotic, highly dynamic digital environment include: "Innovations, Authenticity, Consumer-centricity."

6 Conclusion

The change in brand-consumer **interaction** in the social network era is a result of dynamic play in emergent trends with challenges that prevailed along the progress of branding strategies. Moving ahead through such a constantly changing ecosystem, brands need agility and innovation, along with a consumer-centric focus while negotiating complexities in the digital landscape.

Emerging trends in social commerce, AR experiences, and personalized content are the new dimensions in which brands can innovate customer engagement. Those companies which can follow such patterns through the usage of new technologies inclusive of voice search and AI will be able to provide seamless experiences to customers with much better resonance and bottom-line effectiveness in driving sales.

Social media marketing strategy offer a brand an opportunity to explore the dimensions of engagement and identity building through consumer-oriented dialogues on Twitter, Instagram, and Facebook.

These brand engagements happen through users discussing the site spontaneously; this allows brand messages to get across the networks organically, rather than being communicated artificially from company to consumer.

Social media also extends a strong possibility for brands to emotionally connect with their audience and thereby influence their choice of making a purchase. By collating and analyzing the mountain of data emanating from these **interactions**, a brand can devise better strategies, has longer-term relationships with its customers, and sharpens its marketing tools.

Yet, for all of the gains promised by these budding trends, each faces a multitude of challenges that include, first and foremost, content glut, authenticity questions, and gnarled regulatory ecosystems. In light of these obstacles, it follows logically that brands will have to keep in mind the need for transparency, integrity, and consumer trust while negotiating at speed regulations and platform requirements that are just as suddenly changing.

Of course, digitalization is a set of complex challenges in itself. As much as social media sites are a powerhouse of engagement, at the same time, brands feel pressured to pace up against ever-evolving algorithms, flooding content into the feed, and concerns around privacy. Authenticity is increasingly hard to sustain in this digital sea, where increasingly skeptical and selective consumers make choices about which brands to trust. Furthermore, while technological innovations like **artificial intelligence**, **augmented reality**, and tailored content provide creative avenues of engagement, they require a high degree of adaptability and transparency from the brands in order to deal ethically with consumer data. In this respect, branding has to assume a much more flexible attitude in order to maintain competitiveness and consumer focus. The digital brand, it is postulated in this paper, needs to move with rapid changes in technologies and the evolution of consumers' expectations and balance innovation with core brand values, considerations of privacy, and trust. Digitization has therefore redefined not just how to go about branding but also asks brands to be ever-vigilant and responsive to the vagaries of the digital landscape

The transformation of branding tactics during the age of social networks emphasizes the criticality for companies to adopt innovative approaches, cultivate genuineness, and place customer involvement at the forefront of their online branding endeavors. The further research agenda should try to investigate the impact of emerging technologies like AR and VR on branding, the effect of **artificial intelligence** on **personalization**, and the effectiveness of influencer marketing. Of particular need for research is how to manage the ever-growing **user-generated content**, how brands can fit changes in social media algorithms, and what brands can do to remain true and trusted by consumers in a content-overload world.

This paper summarizes branding in the digital era, with respect to critical strategies and new trends that brands must adopt to be competitive. Taking an in-depth look at these developments, this paper will be useful as a base for understanding the changing scenario in **digital branding** and the methods that enable brands to create lasting and authentic relationships with consumers.

References

Kingsnorth, S.: Digital Marketing Strategy: An Integrated Approach to Online Marketing. Kogan Page (2022)

Schwartz, E.: Product-Led SEO: The Why Behind Building Your Organic Growth Strategy. O'Reilly Media (2021)

Enge, E., Spencer, S., Strichola, J.: The Art of SEO: Mastering Search Engine Optimization. O'Reilly Media (2023)

Baer, J.: Youtility: Why Smart Marketing Is About Help Not Hype. Portfolio/Penguin., New York (2024)

Barnard, J.: The Fundamentals of Brand SERPs for Business. Kogan Page (2022)

Puzakova, M., Aggarwal, P.: Consumer desire for distinctiveness and anthropomorphized brands. J. Consum. Res., Oxf. Acad. **45**(4), 869–888 (2018)

Calder, B.J., Malthouse, E.C., Vandenbosch, M.: Creating brand engagement on digital, social and mobile media. In: Customer Engagement, pp. 85–101. Routledge (2015)

Verhoef, P.C., Reinartz, W.J., Krafft, M.: Consumer engagement as a new perspective in consumer management. J. Serv. Res. **13**(3), 247–252 (2010)

Bijmolt, T., Dorotic, M., Verhoef, P.C.: Loyalty programs: Generalizations on their adoption, effectiveness and design. Found. Trends Mark. **5**(4), 197–258 (2010)

van Doorn, J., et al.: Consumer engagement behaviour: Theoretical foundations and research directions. J. Serv. Res. **13**(3), 253–266 (2010)

Ramaswamy, V., Ozcan, K.: The impact of digital transformations on consumer-brand relationships. Int. J. Res. Mark., Emerald Insight. (2016)

Gambetti, R.C., Biraghi, S., Schultz, D.E., Graffigna, G.: Brand wars: Consumer—Brand engagement beyond client–agency fights. J. Strateg. Mark. **24**(2), 90–103 (2016)

Kim, A.J., Johnson, K.K.: Power of consumers using social media: Examining the influences of brand-related user-generated content on Facebook. Comput. Hum. Behav. **58**, 98–108 (2016)

Shao, Z.: How social media influencers influence consumers' impulsive buying. J. Res. Interact. Mark., Emerald Insight. **124**, 343 (2023)

Hsieh, J.K., Kumar, S.: The effect of consumers' need for touch on webrooming intention. J. Res. Interact. Mark., Emerald Insight. (2023)

Lang, G., Kenny, G.: Strategic approaches to digital branding. J. Bus. Strategy, Emerald Publ. (2024)

Goyal, A., Verma, P.: The relationship between brand engagement, brand loyalty, overall brand equity and purchase intention. J. Strateg. Mark. **32**(1), 65–79 (2024). https://doi.org/10.1080/0965254X.2022.2149839

Gil, C.: The End of Marketing: Humanizing Your Brand in the Age of Social Media and AI. Independent Publisher (2019)

Vasilj, D., Stojkić, I., Bubalo, N.: Artificial intelligence in marketing and public relations of business organizations. South East. Eur. J. Commun. **5**(2), 61–69 (2023)

Vasilj, D., Galić, M., Volarić, T.: Selection of quality staff for jobs in the it sector. In: 23rd International Symposium on Quality—Yesterday, Today, Tomorrow, pp. 525–535 (2022)

Vasilj, D., Vučić, F., Matković, Ž.: Application of artificial intelligence in the economic and legal affairs of companies. In: International Conference on Digital Transformation in Education and Artificial Intelligence Application, pp. 229–240. Springer Nature, Switzerland (2024)

Safeer, A.A.: Harnessing the power of brand social media marketing on consumer online impulse buying intentions: A stimulus-organism-response framework. J. Prod. Brand Manag. **33**(5), 533–544 (2024)

Libai, B., et al.: Customer-to-customer interactions: Broadening the scope of word of mouth research. J. Serv. Res. **13**(3), 267–282 (2010)

Yudhi, O., et al.: The role of social media marketing, entertainment, customization, trendiness, interaction and word-of-mouth on purchase intention: An empirical study from Indonesian smartphone consume. Int. J. Data Netw. Sci. **5**(3), 231–238 (2021)

Bartschat, M., Cziehso, G., Hennig-Thurau, T.: Searching for word of mouth in the digital age: Determinants of consumers' uses of face-to-face information, internet opinion sites, and social media. J. Bus. Res. **141**, 393–409 (2022)

How Consumer Environmental Awareness Influences the Effectiveness of Green Advertising? Evidence from Serbia

Jelena Krstić[1](✉), Nena Vasić[2], and Tamara Vlastelica[3]

[1] Institute of Economic Sciences, 11000 Belgrade, Serbia
jelena.krstic@ien.bg.ac.rs
[2] Mozzart d.o.o, 11000 Belgrade, Serbia
[3] Faculty of Organizational Sciences, University of Belgrade, 11000 Belgrade, Serbia

Abstract. There are multiple consumer-related factors that influence the effectiveness of green marketing, some of which are consumers' pro-environmental values, attitudes and behaviour. The objective of this research was to determine whether consumers' environmental awareness can influence the effectiveness of green advertising, and whether this relationship is moderated by demographic characteristics of consumers. An online survey was used to conduct a survey on a sample of 270 consumers in Serbia. The findings indicate that variables related to consumers' environmental awareness (Pro-environmental concern, Perceived consumer effectiveness, Perception of personal gains, Tendency to conscious consumption and Responsible waste disposal) positively influence variables indicating effectiveness of green marketing (Consumer receptiveness to green advertising, Consumer skepticism towards green advertising and Perceived credibility of eco-labels). The relationship between these two groups of variables was moderated by consumers' education level and monthly income. These findings contribute to the enlarging literature in this field, and represent a valuable basis for the development of effective green marketing and communication strategies in business practice.

Keywords: Consumer Environmental Awareness · Green Advertising · Effectiveness · Demographics · Education Level · Monthly Income

1 Introduction

Environmental problems, such as hazardous waste, soil, water and air pollution, global warming, are critical as they threaten the planet's ecosystems, biodiversity, human health, and the future of sustainable development, making urgent action necessary. Due to that, economic, social and environmental values are gaining more and more importance in today's business environment [1]. Both companies and consumers have become more concerned about the impact of the production and the consumption on the environment. From the corporate perspective, the increasing attention put to the corporate responsibility in the environmental protection, affected their supply and production practices, and this is also reflected in their communications [2, 3]. Thus, if companies tend to achieve

V. Štavljanin et al. (Eds.): SymOrg 2024, LNNS 1681, pp. 62–77, 2025.
https://doi.org/10.1007/978-3-032-08090-5_6

and maintain business and market success, they will have to accept the green practices and modify their business in environmentally responsible way.

On the other side, **consumer environmentally responsible behaviour** is also crucial as it drives demand for sustainable products, reduces environmental impact, and encourages businesses and governments to prioritize eco-friendly practices. Even though consumers are becoming more and more aware and informed about the environmental impact of their consumption, their actual behaviour is still not at the satisfactory level. Therefore, encouraging consumers to adopt more sustainable behaviors is one of today's greatest challenges [4]. Green behaviour is significantly influenced by an individual's **environmental awareness**, which can be enhanced by educating consumers about sustainable choices and the environmental impact of their actions [5]. **Green advertising**, which is conveyed to the public through various media, such as television, radio, print and digital media platforms, is considered to be one of the factors which may influence consumers. Through green advertising, organizations, in addition to promoting an environmentally acceptable products and building green corporate identity, also influence consumer attitudes and behaviour [6].

The effectiveness of environmental advertising thus lies in its potential to attract attention and make an influence on consumers' perceptions, attitudes and behaviour [7]. For such reason, this particular topic has been gaining importance for professionals and academics in the field of marketing, resulting in the identification of various determinants of green advertising effectiveness and recommendations for its improvement. There is a number of studies dealing with different aspects of green advertising effectiveness (e.g [7–9].). Whereas majority of research focuses on the impact of green advertising on consumer attitudes and behaviour (e.g [9, 10].), less attention is paid to the relationship of consumers' environmental awareness and the effectiveness of green advertising. Thus, the objective of this paper is to examine whether there is an impact of the existing level of consumers environmental awareness on the effectiveness of green advertising, and whether this relationship is moderated by consumers demographic characteristics.

This chapter is divided into five sections. After the introductory section, the second section is dedicated to literature review on all concepts included in the research. The third section includes methodology and the fourth section includes presentation of the findings. At the end, there is a concluding section, containing an overview of the obtained results, as well as limitations and future research directions.

2 Literature Review

2.1 Consumer Environmental Awareness

According to Grant [11], the increase in the awareness of the general public and companies about environmental problems, resulted in a change in consumers lifestyle because they have been gradually adopting habits of sustainable and environmentally friendly behavior. Generally speaking, consumers who are characterized as green are aware of the impact of products on health and the environment. Therefore, these consumers demand that products they choose do not harmfully affect the environment during their production, use and disposal, that their production or testing does not involve animal cruelty, and that these products do not affect human health in general. The result of increasing environmental awareness is a shift in consumer behavior towards more environmentally responsible practices [12, 13]. According to Sana [14], consumers who are environmentally conscious demonstrate a tendency to satisfy their needs with non-harmful or at least less harmful products in relation to the environment. By applying green advertising strategies, organizations can enhance the rise of environmental awareness among consumers [15]. According to Chan et al. [16], green awareness itself is the result of marketing efforts invested by organizations, manifested through green advertising and green brand image, and once it is developed, green awareness becomes the antecedent of consumer intentions and purchasing decisions.

From the consumer's point of view, green awareness reflects in a spectrum of different activities, from the reduction of consumption of conventional products, up to recommendations and choices of products that are considered to be more environmentally friendly [17]. Generally, environmental awareness represents a broad concept which includes multiple sub-concepts, among which are: Pro-environmental concern, perceived consumer effectiveness, perception of personal gains, tendency to conscious consumption and responsible waste disposal.

Pro-environmental Concern (PEC). Increasing concern for the severity of issues like environmental degradation and climate change has highlighted the importance of exploring social and behavioral aspects of environmental problems more than ever before [18]. As defined by Fransson and Gärling [19], environmental concern ranges "from a specific attitude toward environmentally relevant behaviour to a more encompassing value orientation". Despite the representation of these topics in public discourse, there is considerable variety in consumers' attitudes toward environmental issues, so while some of them observe the environment from a purely utilitarian point of view, others are highly concerned about solving environmental problems and achieving sustainability [20]. Since pro-environmental concern is influenced by numerous personal and social factors [21], it can be concluded that consumers might demonstrate different levels of concern depending on their demographics, cultural and economic background and similar.

Perceived Consumer Effectiveness (PCE). As widely accepted, perceived consumer effectiveness represents the degree up to which consumers perceive they can, at the individual level, influence the marketplace and contribute [22–24]. Perceived consumer effectiveness has proved to be one of the determinants of pro-environmental behaviour [22, 23], meaning that it is more likely that consumers with higher levels of perceived consumer effectiveness would behave in a more environmentally friendly manner [25].

Perception of Personal Gains (PPG). Perception of gains from certain behaviour represents significant motivational factor to behave in a certain way and is found to be effective in an environmentally oriented communication [26]. Palacios-González and Chamorro-Mera [27] explained this concept as "the individual's subjective assessment of the advantages and disadvantages of socially responsible behavior". As explained by these authors, this subjective assessment can be expressed in a positive sense, pointing to the expected gains, or in a negative sense, reflecting feelings of being harmed in certain way (e.g. such as spending more time on making purchasing decision or paying higher prices). In this research, the perception of personal gains from environmentally responsible consumption can be explained as consumers' belief that such behaviors not only benefit the environment but also enhance one's well-being.

Tendency to Conscious Consumption (TCC). According to Yan and She [28], responsible consumption reflects a set of actions taken throughout the consumption process, shaped by consumers concern for social, ethical, and environmental issues. Gupta and Agrawal [29] defined conscious consumption as the extent to which individual's consumption is in line with individual's real needs and tendency to reduce waste. Vlastelica et al. [30] explained the conscious consumption "as a frugal behavior towards natural resources, which is connected with consumers' caring of the environmental well-being". Therefore, consumers' tendency to conscious consumption can be defined as the deliberate choice to purchase and use products or services that align with their ethical, social, and environmental values.

Responsible Waste Disposal (RWD). Improper waste disposal represents signific threat to the environment and human health [31]. In contrast, responsible waste management enables the conservation of energy and resources and provides clear support to the preservation of the environment [32]. Due to the significance of efficient practices in this field, there is growing research exploring drivers of consumers' willingness to dispose of used products in a socially and environmentally responsible way [33]. Waste disposal which is observed as responsible includes actions such as waste segregation, proper handling of harmful and hazardous waste, recycling, etc. [29].

2.2 Green Advertising Effectiveness

In today's social and market landscape, environmental awareness and eco-friendly consumption play a crucial role in shaping consumer behavior. Organizations invest significant resources in the development of green products and green marketing communications to build and maintain desirable reputation among consumers and align their offerings with global environmental trends. Effective communication is crucial because it ensures that conveyed messages are clearly understood, which supports consumer decision-making and fosters trust and developing relationships [34–36]. Green communication includes various forms of communication, such as green advertising and eco-labelling. Green advertising focuses on delivering messages that raise awareness of environmental issues, promote eco-friendly products, and encourage behaviors that help reduce negative environmental impact [37]. By one of the earlier definitions, it is defined as a type of promotional message that is in line with the needs and demands of environmentally conscious consumers [38]. Goals of green advertising refer to the efficient informing of consumers regarding environmental aspects of the organization's products and services, building awareness and positive attitudes towards environmentally friendly products [39–41]. Eco-labels, representing a form of green communications, serve as certification marks or seals of approval, indicating that product or service possess certain environmental attributes and simultaneously assures consumers that these claims are accurate [43]. Whereas green advertising has proved to be highly effective in influencing the improvement of environmental awareness among consumers, some of the previous research indicated that the effectiveness might depend on consumers' current environmental consciousness level [44]. Depending on their environmental awareness, consumers might be unequally receptive to green advertising, express different levels of skepticism towards green advertisements and differently perceive the credibility of eco-labels.

Consumer Receptiveness to Green Advertising (RGA). This concept pertains to the degree of consumer attention and positive responsiveness to advertisements that feature environmentally friendly messages, whether these ads promote specific products or the company as a whole. Green advertising research has identified its critical role in creating favorable consumer perceptions [45] and promoting environmentally conscious consumption behavior [46]. The aspiration of many organizations is to make green information related to their products and services credible [47, 48], so consumers will perceive it in a positive way. Green advertisements are known to use a range of appeals, to help enhance the persuasiveness of green advertisements [49, 50]. Specific, relevant and complete information provided in advertisements can improve the effectiveness of communication and increase the likelihood that consumers will trust green advertisements [43]. Some of the research findings indicate that consumers who have already developed positive attitudes towards the preservation of the environment and thus have high environmental awareness, are more likely to react to green communication in a positive way [17].

Consumer Skepticism Towards Green Advertising (SGA). This concept reflects consumers' potential reactions to perceived exaggerations, misleading or confusing information, and the perceived accuracy of environmental claims [51]. The notion of the skeptical green consumer in literature related to those consumers who especially consider green advertisements to be false and misleading. In order to be effective, green claims should be authentic and trustful, conveying a sincere desire of a company to make their business environmentally responsible [52]. Otherwise, it they are not considered to be authentic and well argued, green advertisements are less likely to be effective and instead they can evoke consumers' skepticism towards green messages [45]. Observing the informational content of advertising claims, numerous studies have judged green ads to be ambiguous, confusing or even misleading to consumers [47, 49, 53]. The rise of consumer skepticism towards advertising can undermine trust in green advertising and influence how consumers interpret the motives of advertisers promoting environmentally-focused messages, which consequently reduces the effectiveness of green advertising. This occurs when consumers perceive that certain claims may be deceptive and indicating green washing [42, 54].

Perceived Credibility of Eco-Labels (CEL). Eco-labels are information tools that "aim to internalize the external effects of the production, consumption and disposal of products on the environment" [55]. Concrete arguments consistently enhance greater trust in the eco-label and strengthen positive attitudes towards the product and label source [43]. The findings of Cai et al. [56], indicated that higher levels of consumers' perception of the credibility of eco-labels was connected with higher purchasing intentions of green products. The more consumers feel trustful towards an eco-label, the more positively it would affect their use of that label as an information cue in purchase decisions [57].

Based on the literature review and previous research findings in this area, the following research questions can be set:

RQ1: Are variables of consumer environmental awareness (pro-environmental concern, perceived consumer effectiveness, perception of personal gains, tendency to conscious consumption and responsible waste disposal) significantly related to variables of green advertising effectiveness (consumer receptiveness to green advertising, consumer skepticism towards green advertising, perceived credibility of eco-labels)?

RQ2: Are these relationships moderated by consumers' demographic characteristics?

3 Methodology

The research was carried out on a convenient sample of 270 Serbian consumers between November 2022 and February 2023. Data was collected through an online survey. As regarding sample structure, there were more female than male participants (57% vs. 43%). By observing the age structure, it can be seen that the largest percentage of respondents was younger than 30, even though the percentage of those between 30 and 50 was not negligible. Additionally, as much as 80% of respondents has gained university education. Also, employed respondents made up the majority (87.4%) of the sample (Table 1).

The questionnaire was divided into several segments, with the first segment focusing on the respondents' demographic characteristics. Five variables belonging to the segment Consumers' environmental awareness (CEA) were included in the second segment of the questionnaire. This segment consisted of 22 statements where the environmental awareness of consumers was assessed depending on the extent to which they agree with the stated statements. The aforementioned claims indicate the consumer's habits towards environmental protection, the individual's impact on the environment, the socially responsible behavior of the individual, the extent to which consumption is aligned with real needs and waste reduction, and the environmental effects of certain product use and recycling. Three items borrowed from GREEN scale [58] and two from NEP scale [59] constructed the variable pro-environmental concern (PEC). The following variable, which included three items adopted from Roberts [60], was entitled perceived consumer effectiveness (PCE). The third variable was entitled perception of personal gains and it consisted of items adopted from Palacios-González and Chamorro-Mera [27]. Both variables called Tendency to conscious consumption (TCC) and Responsible waste disposal (RWD) included items borrowed from earlier study of Gupta and Agrawal [29].

The third section of the questionnaire included three variables related to **green advertising effectiveness**. By evaluating the above 17 claims, the respondents pointed to their reactions to environmentally oriented communication, skepticism towards green claims and the trust they have in ecological labels. The first, labelled as Consumer receptiveness to green advertising (RGA) included four items from REGRAD scale [61]. Consumer skepticism towards green advertising (SGA) was constructed of three items previously defined by Do Paço and Reis [62]. The last variable, entitled Perceived credibility of eco-labels (CEL), included items adopted from the study of Taufique et al. [63]. The extent of respondents' agreement with the items related to all variables was expressed

Table 1. Demographic characteristics of the sample.

Demographics	Options	N (%)
Gender	Male	116 (43.0)
	Female	154 (57.0)
Age	< 30	134 (49.6)
	30–50	122 (45.2)
	> 50	14 (5.2)
Education	Secondary	26 (9.6)
	College	26 (9.6)
	University	218 (80.7)
Employment status	Employed	236 (87.4)
	Unemployed	34 (12.6)
Household type	Single	70 (25.9)
	Partnership/marriage without children	54 (20.0)
	Partnership/marriage with children	72 (26.7)
	Living with parents/ siblings	60 (22.2)
	Roommates	14 (5.2)
Monthly income	Below average	94 (34.8)
	Average	70 (25.9)
	Above average	106 (39.3)

on a scale ranging from 1 to 4, where 1 meant "I do not agree at all" and 4 "I completely agree". In the last segment, respondents had to choose factors which they consider to be the most influential on their decision to purchase an ecological product/service. Respondents had the opportunity to choose several options among those offered on the list. For the purpose of data analysis, based on the methods of descriptive statistics, correlation and regression analysis, the statistical software SPSS 25 was used.

4 Results

The descriptive statistical analysis of all variables belonging to CEA and GME was conducted as the initial phase of the analysis. Taking into account that mean values of all variables exceeded the value 2, it can be concluded that respondents' assessment of the proposed items was, in general, favorable (Table 2). Within the group of CEA variables, TCC was the variable with the highest mean value, having a score of 3.17, whereas PCE was the variable with the lowest mean value of 2.26. This suggests that while respondents recognized the proper use of products to protect the environment as highly important, they still believed their individual roles in environmental protection were not particularly significant. Among the variables belonging to the GME group, CEL was the variable with the highest mean value of item ratings (3.08), while SGA was the variable with the lowest ratings. This indicates that respondents genuinely believed in the environmental claims made by brands they trust. Furthermore, a Cronbach's Alpha coefficient was determined for each variable separately in order to determine the reliability of the analyzed variables. Taking into account that values of this coefficient ranged from 0.510 to 0.871, it can be concluded that the survey instrument had satisfactory measurement characteristics.

Table 2. Mean values of variables.

Group	Variable	Mean (± SD)
Consumer environmental awareness (CEA)	PEC	2.83 (±0.582)
	PCE	2.26 (±0.649)
	PPG	2.36 (±0.822)
	TCC	3.17 (±0.754)
	RWD	2.36 (±0.762)
Green marketing effectiveness (GME)	RGA	2.87 (±0.807)
	SGA	2.35 (±0.809)
	CEL	3.08 (±0.668)

As the following step, the mutual relations between the variables from both groups, CEA and GME, were analyzed by applying a correlation analysis. The results of this analysis revealed a positive and statistically significant correlation between both groups of variables (Table 3).

Table 3. Correlation between CEA and GME variables.

	PEC	PCE	PPG	TCC	RWD
RGA	0.622[a]	0.216[a]	0.180[a]	0.463[a]	0.539[a]
SGA	0.328[a]	0.500[a]	0.414[a]	0.206[a]	0.400[a]
CEL	0.503[a]	0.287[a]	0.215[a]	0.515[a]	0.460[a]

a The correlation is significant at the level of 0.01.

The achieved values of the correlation between the analyzed variables belonging to CEA and GME were statistically significant and indicated that there was a moderate to significant correlation. All values achieved positive values of low to moderate correlation, which confirmed the relation between consumers environmental awareness and effectiveness of green marketing. The correlation appeared to be the highest between variables PEC and RGA and RWD and RGA. This can be interpreted in a way that consumers who had already adopted environmentally friendly habits and avoided using products that could negatively impact the environment seemed more receptive to green marketing. Likewise, when they understood the importance of proper waste disposal, they showed a greater responsiveness to green advertising and other marketing communication forms. Additionally, variables PEC and TCC were significantly correlated with the variable CEL. This means that consumers who expressed higher pro-environmental concern and mindfulness of their waste disposal practices also perceived the credibility of eco-labels better. On the other hand, variables PPG and RGA and variables TCC and SGA had the lowest degree of correlation. This suggests that respondents who did not observe personal gains from the environmental preservation also were less positively responsive to environmental messages in marketing communication.

The third step of the analysis included regression analysis, as a means to explore the influence of the variables belonging to CEA on the variables belonging to GME group (Table 4). The achieved R^2 values showed statistical significance. The findings indicated that CEA variables significantly influenced RGA ($R^2 = 0.471$, F change - 46.966, $p < 0.000$), SGA ($R^2 = 0.335$, F change - 26.546, $p < 0.000$) and CEL ($R^2 = 0.366$, F change - 30.469, $p < 0.000$). It can be seen that CEA variables demonstrated highest influence on RGA (R2 = 0.471). Based on this, it can be concluded that the development of environmental awareness among consumers explained 47.1% of consumer receptiveness to green marketing.

Table 4. Regression model of the influence of variables belonging to CEA group on variables belonging to GME.

Dependents	Independents	β	t	Sig.	R^2	F	Sig.
RGA	PEC	0.463	7.918	0.000	0.471	46.966	0.000
	PCE	−0.108	−1.928	0.055			
	PPG	−0.021	−0.387	0.699			
	TCC	0.128	2.336	0.020			
	RWD	0.279	5.044	0.000			
SGA	PEC	0.036	0.544	0.587	0.335	26.546	0.000
	PCE	0.326	5.171	0.000			
	PPG	0.177	2.952	0.003			
	TCC	−0.076	−1,246	0.214			
	RWD	0.257	4.139	0.000			
CEL	PEC	0.244	3.813	0.000	0.366	30.469	0.000
	PCE	0.024	0.392	0.695			
	PPG	0.007	0.124	0.902			
	TCC	0.292	4.869	0.000			
	RWD	0.183	3.021	0.003			

Based on the results of regression analysis, the following influence of the independent variables from CEA group on the dependent variables of GME was determined.

The results of the regression analysis indicated that, in general, the highest statistical significance was observed with RWD, as its independent impact on all analyzed GME variables was confirmed. Variables PEC and RWD appeared to have the greatest individual influence on RGA. Simply put, the results indicated that consumers' pro-environmental concern and tendency towards responsible waste disposal had a positive and significant impact on consumers' responsiveness to green marketing, as a concept related to GME. Furthermore, variables PCE, PPG and RWD appeared to significantly and positively influence consumer skepticism towards green marketing. When it comes to CEL, as one of the components of green marketing effectiveness, the results showed the statistically significant impact of PEC, TCC and RWD. This means that consumer concern for the environmental protection, tendency to consume products according to real needs and to adequately dispose of products positively influenced consumers' perception of the credibility of eco-labels.

In the following stage of the analysis, the relationship of the CEA variables and GME variables was examined with the moderating effect of demographic characteristics. The significant moderating effect was identified in case of education level and household's monthly income. The presented results of the correlation analysis showed that there were significant statistical differences between the established correlation relationships among respondents with different levels of education (Table 5). Namely, as the level of consumer education increased, so does the statistical significance of the correlation relations between CEA variables and GME variables. Among respondents who indicated that they had a secondary education, the variables PCE and PPG did not establish statistically significant correlation values with any of the analyzed variables of GME except one. When it comes to the respondents with college education, the negative correlation between PPG and RGA indicated that, among this demographic group, respondents did not have enough awareness of the importance of responsible behavior towards the environment and the benefits that can result from it.

In case of respondents with university education, the correlation between variables showed positive and statistically significant values. This indicates that this group of consumers had developed awareness of environmentally responsible and the role of the individual in it, and certainly that this group of respondents expressed greater confidence in ecological claims and promotional messages. On the basis of these results, it was confirmed that variables related to consumer environmental awareness demonstrated statistically significant relationships with variables related to effectiveness of green marketing, with the moderating effect of the level of respondents' education.

As presented in Table 6, the correlation between CEA and GME variables also was moderated by the level of households' monthly income. What is noticeable is that, with the increase of the monthly income, the intensity and strength of correlations between the analyzed variables also increased. In the case of consumers who stated that they had below average income, it was determined that certain correlation values did not have statistical significance and that negative correlations appeared. Namely, the variables of PCE and PPG had negative correlations with variables RGA and CEL. Therefore, the inability of these respondents to allocate more funds for the purchase of green products creates dissatisfaction and mistrust among these consumers that they, as individuals, can

Table 5. Correlation between CEA and GME variables with the moderating effect of respondents' level of education.

Education		PEC	PCE	PPG	TTC	RWD
Secondary	RGA	0.803**	0.039	0.369	0.594**	0.895**
	SGA	409*	0.186	0.259	0.614**	0.234
	CEL	0.666**	0.363	0.437*	0.828**	0.658**
	N	26	26	26	26	26
College	RGA	0.531**	0.177**	−0.006	0.434**	447**
	SGA	0.282**	0.486**	0.390**	0.123	0.417**
	CEL	0.399**	0.237**	0.150*	0.446**	0.395**
	N	26	26	26	26	26
University	RGA	0.825**	0.475*	0.170*	0.545**	0.707**
	SGA	0.390	0.670**	0.613**	0.487*	0.434*
	CEL	0.893**	0.519**	0.360	512**	0.610**
	N	218	218	218	218	218

a The correlation is significant at the level of 0.01
b The correlation is significant at the level of 0.05

influence the preservation of the environment by using such products and can benefit from such behaviour.

Correlation values for consumers who stated that they have an average and above average monthly income indicated positive and statistically significant values. Therefore, on the basis of the obtained results, it can be concluded that variables from the consumers environmental awareness group showed statistically significant relationships with the variables from the effectiveness of green marketing group, with the moderating effect of the household's monthly income.

In the final segment of the analysis, the significance of different factors on respondents' decision to choose green product/service was assessed based on the obtained responses.

As can be seen in Table 7, the largest percentage of respondents (21.7%) stated that, when purchasing an environmentally friendly product, they are guided by advertising. A slightly smaller number of respondents (17.1%) cited that their personal responsibility for the environment influences their decision to purchase environmentally friendly product/service. Recognizable brand (corporate/product) was noted to be influential factor by 12.5% of respondents. A small number of respondents (0.3%) did not use environmentally friendly products or did not have an opinion on this topic.

Table 6. Correlation between CEA and GME variables with the moderating effect of the amount of the household's monthly income.

Monthly income		PEC	PCE	PPG	TCC	RWD
Below average	RGA	693[a]	−0.055	−0.067	0.462[a]	0.619[a]
	SGA	0.155	0.357[a]	0.460[a]	0.104	0.437[a]
	CEL	0.470[a]	0.027	−0.097	0.430[a]	0.365[a]
	N	94	94	94	94	94
Average	RGA	0.475[a]	0.319[a]	0.341[b]	0.371[b]	0.583[a]
	SGA	471[a]	0.579[a]	302[a]	322[a]	0.353[a]
	CEL	449[a]	0.377[a]	427[a]	495[a]	0.519**
	N	70	70	70	70	70
Above average	RGA	0.700[a]	0.430[a]	322[a]	595[a]	0.383[a]
	SGA	0.421[a]	0.608[a]	525[a]	0.184	414[a]
	CEL	0.611[a]	459[a]	353[a]	644[a]	0.506[a]
	N	106	106	106	106	106

a The correlation is significant at the level of 0.01.
b The correlation is significant at the level of 0.05.

Table 7. Factors which influence respondents' decision to purchase a green product/service.

Factors	N (%)
Personal responsibility for the environment	108 (17.1)
Sales promotion activities	28 (4.4)
Advertising	137 (21.7)
Recognizable brand (corporate/product)	79 (12.5)
Dissatisfaction with purchasing non-ecological alternatives	41 (6.5)
Price (at least equal to the price of the non-ecological variant)	47 (7.4)
Label "eco" or "organic" on the packaging	53 (8.4)
Chemical composition of the product	58 (9.2)
Packaging	67 (10.6)
Social influence	10 (1.6)
I do not use green products	2 (0.3)
I don't know	2 (0.3)

5 Conclusion

Many organizations are including forms of green advertising in their marketing strategies and this concept involves distributing messages which promote awareness of environmental issues, benefits from purchasing green products/services or suggest certain behavioral changes that can be purposeful for reducing the negative impact of overconsumption on the environment. Consumers are increasingly encouraged to behave in an environmentally responsible way, and therefore to opt for green products and adopt green practices as much as possible. However, it was notified in literature and marketing practice that the effectiveness of green advertisements might vary depending on personal characteristics of consumers. Based on the results of correlation and regression analysis in this study, the existence of statistically significant relationship between consumer environmental awareness and green advertising effectiveness was found. In particular, it was found that variables indicating the level of environmental awareness of consumers (Pro-environmental concern, Perceived consumer effectiveness, Perception of personal gains, Tendency to conscious consumption and Responsible waste disposal) influence variable indicating the effectiveness of green advertising (Consumer receptiveness to green advertising, Consumer skepticism towards green advertising and Perceived credibility of eco-labels). Therefore, it can be concluded that the higher environmental awareness of consumers was, more positively they would react to green messages and such messages would have greater influence on them. Also, it was found that the level of education and household's monthly income were two characteristics which affect the relationship between these two concepts.

The managerial implications of the study indicate that managers should tailor green marketing strategies to align with the environmental awareness of their target audience, emphasizing credible eco-labels and reducing skepticism through transparent communication. Efforts should be focused on segments with higher education and income levels, as these consumers are more receptive to green marketing.

The results of this research contribute to the growing literature in the field of green marketing and consumer behavior, especially when it comes to empirical research in developing countries where the environmental awareness of consumers is still not at a satisfactory level. On the other hand, the research also has certain shortcomings. The research did not include a representative sample, which reduces the capacity to generalize the results. Also, the included variables are the result of the author's subjective assessment of their suitability, so it is possible that the choice of other variables related to the two key concepts (environmental awareness and advertising effectiveness) would lead to different results.

References

1. Dahlstrom, R.: Green Marketing Management. South-Western Cengage Learning, Mason, OH (2011)
2. Slijepčević, M., Popović Šević, N., Krstić, J., Rajić, T., Ranković, M.: Exploring the nexus of perceived organizational CSR engagement, job satisfaction, organizational pride, and involvement in CSR activities: Evidence from an emerging economy. Sustainability. **16**, 3403 (2024)

3. Vlastelica, T., Kostic, S.C., Okanovic, M., Milosavljevic, M.: How corporate social responsibility affects corporate reputation: Evidence from an emerging market. J. East Eur. Manag. Stud. **23**(1), 10–29 (2018)
4. White, K., Habib, R., Hardisty, D.J.: How to SHIFT consumer behaviors to be more sustainable: A literature review and guiding framework. J. Mark. **83**(3), 22–49 (2019)
5. Usrey, B., Palihawadana, D., Saridakis, C., Theotokis, A.: How downplaying product greenness affects performance evaluations: Examining the effects of implicit and explicit green signals in advertising. J. Advert. **49**(2), 125–140 (2020)
6. Kong, Y., Zhang, A.: Consumer response to green advertising: The influence of product involvement. Asian J. Commun. **23**(4), 428–447 (2013)
7. Shen, W., et al.: The impact of advertising creativity, warning-based appeals and green dispositions on the attentional effectiveness of environmental advertisements. J. Clean. Prod. **271**, 122618 (2020)
8. Royne, M.B., Martinez, J., Oakley, J., Fox, A.K.: The effectiveness of benefit type and price endings in green advertising. J. Advert. **41**(4), 85–102 (2012)
9. Shin, S., Ki, E.J., Griffin, W.G.: The effectiveness of fear appeals in 'green'advertising: An analysis of creative, consumer, and source variables. J. Mark. Commun. **23**(5), 473–492 (2017)
10. Alniacik, U., Yilmaz, C.: The effectiveness of green advertising: Influences of claim specificity, product's environmental relevance and consumers' pro-environmental orientation. Amfiteatru Econ. J. **14**(31), 207–222 (2012)
11. Grant, J.: The Green Marketing Manifesto. John Wiley & Sons, Hoboken, NJ (2009)
12. Chang, H., Zhang, L., Xie, G.X.: Message framing in green advertising: The effect of construal level and consumer environmental concern. Int. J. Advert. **34**(1), 158–176 (2015)
13. Kanonuhwa, M., Chimucheka, T.: Green marketing and purchase behaviour of generation Y-consumers. Mediterr. J. Soc. Sci. **5**(20), 2785 (2014)
14. Sana, S.S.: Price competition between green and nongreen products under corporate social responsible firm. J. Retail. Consum. Serv. **55**, 102–118 (2020)
15. Rahmi, D.Y., Rozalia, Y., Chan, D.N., Anira, Q., Lita, R.P.: Green brand image relation model, green awareness, green advertisement, and ecological knowledge as competitive advantage in improving green purchase intention and green purchase behavior on creative industry products. J. Econ. Bus. Account. Ventura. **20**(2), 1126 (2017)
16. Chan, H.K., He, H., Wang, W.Y.C.: Green marketing and its impact on supply chain management in industrial markets. Ind. Mark. Manag. **41**(4), 557–562 (2012)
17. Suki, N.M.: Green awareness effects on consumers' purchasing decision: Some insights from Malaysia. Int. J. Asia-Pac. Stud. **9**(2), 49–63 (2013)
18. Cruz, S.M., Manata, B.: Measurement of environmental concern: A review and analysis. Front. Psychol. **11**, 363 (2020)
19. Fransson, N., Gärling, T.: Environmental concern: Conceptual definitions, measurement methods, and research findings. J. Environ. Psychol. **19**(4), 369–382 (1999)
20. Hirsh, J.B.: Personality and environmental concern. J. Environ. Psychol. **30**(2), 245–248 (2020)
21. Gifford, R., Nilsson, A.: Personal and social factors that influence pro-environmental concern and behaviour: A review. Int. J. Psychol. **49**(3), 141–157 (2014)
22. Lee, Y.-K., Kim, S., Kim, M.-S., Choi, J.-G.: Antecedents and interrelationships of three types of pro-environmental behavior. J. Bus. Res. **67**(10), 2097–2105 (2014)
23. Higueras-Castillo, E., Liébana-Cabanillas, F.J., Muñoz-Leiva, F., García-Maroto, I.: Evaluating consumer attitudes toward electromobility and the moderating effect of perceived consumer effectiveness. J. Retail. Consum. Serv. **51**, 387–398 (2019)

24. Hanss, D., Doran, R.: Perceived consumer effectiveness. In: Leal Filho, W., Azul, A.M., Brandli, L., Özuyar, P.G., Wall, T. (eds.) Responsible Consumption and Production. Encyclopedia of the UN Sustainable Development Goals. Springer, Cham (2020)
25. Riesgo, S.B., Lavanga, M., Codina, N.: Drivers and barriers for sustainable fashion consumption in Spain: A comparison between sustainable and non-sustainable consumers. Int. J. Fash. Des. Technol. Educ. **16**(1), 1–13 (2023)
26. Sacchi, S., Stanca, L.: Asymmetric perception of gains versus non-losses and losses versus non-gains: The causal role of regulatory focus. Behav. Decis. Mak. **27**(1), 48–56 (2013)
27. Palacios-González, M.M., Chamorro-Mera, A.: Analysis of socially responsible consumption: A segmentation of Spanish consumers. Sustainability. **12**(20), 8418 (2020)
28. Yan, J., She, Q.: Developing a trichotomy model to measure socially responsible behaviour in China. Int. J. Mark. Res. **53**(2), 253–274 (2011)
29. Gupta, S., Agrawal, R.: Environmentally responsible consumption: Construct definition, scale development, and validation. Corp. Soc. Responsib. Environ. Manag. **25**(4), 523–536 (2018)
30. Vlastelica, T., Kostić-Stanković, M., Rajić, T., Krstić, J., Obradović, T.: Determinants of young adult consumers' environmentally and socially responsible apparel consumption. Sustainability. **15**, 1057 (2023)
31. Tlou, R., Nelisiwe, M., Mariana, E.: The impact of improper waste disposal on human health and the environment: A case of Umgungundlovu District in KwaZulu Natal Province, South Africa. Front. Sustain. **5**, 47 (2024)
32. Wilmott, S., Pasdeki-Clewer, E., Duane, B.: Responsible waste management: Using resources efficiently. Br. Dent. J. **235**, 393–397 (2023)
33. Vlastelica, T., Kostić-Stanković, M., Krstić, J., Rajić, T.: Generation Z's intentions towards sustainable clothing disposal: Extending the theory of planned behavior. Pol. J. Environ. Stud. **32**(3), 2345–2360 (2023)
34. Krstić, J.: Integrated advertising—A modern concept and a new trend in marketing communication. Economic Institute, Belgrade (2020) (In Serbian)
35. Krstić, J., Ćorić, N.: Public health communication: Skin cancer prevention implications. Manag.: J. Sustain. Bus. Manag. Solut. Emerg. Econ. **29**(1), 63–85 (2021)
36. Skočajić, I., Pavković, V., Krstić, J., Kostić-Stanković, M., Vlastelica, T., Rajić, T.: Urban tourism appeal: The effectiveness of communication channels and multimedia formats in creating tourists' expectations. Sustainability. **16**, 4552 (2024)
37. Kitić, B., Kostić-Stanković, M., Cvijović, J., Lečić-Cvetković, D.: Environmental aspect of business communications. Manag.: J. Sustain. Bus. Manag. Solut. Emerg. Econ. **20**(74), 69–74 (2015)
38. Zinkhan, G.M., Carlson, L.: Green advertising and the reluctant consumer. J. Advert. **24**(2), 1–6 (1995)
39. D'souza, C., Taghian, M.: Green advertising effects on attitude and choice of advertising themes. Asia Pac. J. Mark. Logist. **17**(3), 51–66 (2005)
40. Ariffin, S., Yusof, J.M., Putit, L., Shah, M.I.A.: Factors influencing perceived quality and repurchase intention towards green products. Procedia Econ. Finance. **37**(16), 391–396 (2016)
41. Green, T., Peloza, J.: Finding the right shade of green: The effect of advertising appeal type on environmentally friendly consumption. J. Advert. **43**(2), 128–141 (2014)
42. Fernandes, J., Segev, S., Leopold, J.K.: When consumers learn to spot deception in advertising: Testing a literacy intervention to combat greenwashing. Int. J. Advert. **39**(7), 1115–1149 (2020)
43. Atkinson, L., Rosenthal, S.: Signaling the green sell: The influence of eco-label source, argument specificity, and product involvement on consumer trust. J. Advert. **43**(1), 33–45 (2014)

44. Bailey, A.A., Mishra, A.S., Tiamiyu, M.F.: Application of GREEN scale to understanding US consumer response to green marketing communications. Psychol. Mark. **35**(11), 863–875 (2018)
45. Matthes, J., Wonneberger, A.: The skeptical green consumer revisited: Testing the relationship between green consumerism and skepticism toward advertising. J. Advert. **43**(2), 115–127 (2014)
46. De Pelsmacker, P., Janssens, W.: A model for fair trade buying behaviour: The role of perceived quantity and quality of information and of product-specific attitudes. J. Bus. Ethics. **75**, 361–380 (2007)
47. Matthes, J., Wonneberger, A., Schmuck, D.: Consumers' green involvement and the persuasive effects of emotional versus functional ads. J. Bus. Res. **67**(9), 1885–1893 (2014)
48. Ganz, B., Grimes, A.: How claim specificity can improve claim credibility in green advertising: Measures that can boost outcomes from environmental product claims. J. Advert. Res. **58**(4), 476–486 (2018)
49. Bickart, B.A., Ruth, J.A.: Green eco-seals and advertising persuasion. J. Advert. **41**(4), 51–67 (2012)
50. Kumar, P.: Intents of green advertisements. Asia Pac. J. Mark. Logist. **29**(1), 70–79 (2017)
51. Mohr, L.A., Eroglu, D., Ellen, P.S.: The development and testing of a measure of skepticism towards environmental claims in marketers' communications. J. Consum. Aff. **32**(1), 30–55 (1998)
52. Leonidou, L.C., Leonidou, C.N., Palihawadana, D., Hultman, M.: Evaluating the green advertising practices of international firms: A trend analysis. Int. Mark. Rev. **28**(1), 6–33 (2011)
53. Fowler, A.R., Close, A.G.: It ain't easy being green: Macro, meso, and micro green advertising agendas. J. Advert. **41**(4), 119–132 (2012)
54. Segev, S., Fernandes, J., Hong, C.: Is your product really green? A content analysis to reassess green advertising. J. Advert. **45**(1), 85–93 (2016)
55. Bougherara, D., Combris, P.: Eco-labelled food products: what are consumers paying for? Eur. Rev. Agric. Econ. **36**(3), 321–341 (2009)
56. Cai, Z., Xie, Y., Aguilar, F.X.: Eco-label credibility and retailer effects on green product purchasing intentions. Forest Policy Econ. **80**, 200–208 (2017)
57. Gorton, M., Tocco, B., Yeh, C.-H., Hartmann, M.: What determines consumers' use of eco-labels? Taking a close look at label trust. Ecol. Econ. **189**, 107173 (2021)
58. Haws, K.L., Winterich, K.P., Naylor, R.W.: Seeing the world through GREEN-tinted glasses: Green consumption values and responses to environmentally friendly products. J. Consum. Psychol. **24**(3), 336–354 (2014)
59. Dunlap, R.E., Van Liere, K.D., Mertig, A.G., Jones, R.E.: Measuring endorsement of the new ecological paradigm: A revised NEP scale. J. Soc. Issues. **56**(3), 425–442 (2000)
60. Roberts, J.A.: Green consumers in the 1990s: Profile and implications for advertising. J. Bus. Res. **36**(3), 217–231 (1996)
61. Bailey, A.A., Mishra, A., Tiamiyu, M.F.: Green advertising receptivity: An initial scale development process. J. Mark. Commun. **22**(3), 327–345 (2016)
62. Do Paço, A.M.F., Reis, R.: Factors affecting skepticism toward green advertising. J. Advert. **41**(4), 147–155 (2012)
63. Taufique, K.M.R., Polonsky, M.J., Vocino, A., Siwar, C.: Measuring consumer understanding and perception of eco-labelling: Item selection and scale validation. Int. J. Consum. Stud. **43**, 298–314 (2019)

Human-Centred Digital Marketing: Emerging Trends and Practical Applications

Katarina Milosavljević(✉), Milica Kostić-Stanković, and Vesna Damnjanović

Faculty of Organizational Sciences, University of Belgrade, Jove Ilića 154, 11010 Belgrade, Serbia
km20225005@student.fon.bg.ac.rs

Abstract. Growth and success are increasingly being driven by digital marketing, especially for major businesses. When creating digital strategies, human elements are being given more and more consideration, which results in more user-centred methods. Products made to meet the needs of people are a clear example of this trend, which is part of a larger wave of innovation in the digital space.

The motivation behind this research stems from the curiosity regarding the extent to which the academic community acknowledges the significance of human-centred approaches in digital marketing studies. The aim is to provide a textual and visual point of view considering the modernist resemblance of the scrutiny of research on human-centred digital marketing and to extract useful academic, practical, and policy implications.

The methodology employed a hybrid bibliometric approach. Initially, a bibliometric analysis was conducted on a corpus of N = 542 papers, which was subsequently narrowed down through a rigorous scrutinisation process to 29 key papers that underwent a systematic literature review. The outcome is a table of research findings and implications for academia, practitioners, and policymakers.

The findings suggest that the academic community should further investigate the application of **augmented reality** (AR) in human-centred digital marketing. Moreover, extended marketing in the era of Marketing 5.0 can enhance the customer shopping experience. Policymakers must consider these behavioural insights when shaping data protection regulations.

Keywords: end-user preferences · human-centred design · human-centred digital marketing

1 Introduction

Digital user engagement has grown dramatically during the last few years, as seen by the proliferation of online information and business transactions. As Patricio *et al.* (2022) said, this change emphasises the need to improve user experience and customise product offers to suit individual tastes better. According to Štavljanin and Pantović (2017), **end-user preferences** are now crucial in redefining product placement strategies and driving ongoing optimisation and refinement. As stated by Kostić-Stanković *et al.* (2020), a two-way communication strategy improves the bond between service providers and customers, which eventually helps businesses expand and get a stronger position in the market.

V. Štavljanin et al. (Eds.): SymOrg 2024, LNNS 1681, pp. 78–92, 2025.
https://doi.org/10.1007/978-3-032-08090-5_7

Promoting goods and services via digital platforms, mobile devices, and other digital technologies—mostly the Internet—is known as **digital marketing** (Visser *et al.*, 2019; Tandava *et al.*, 2022). Global investment in digital marketing is predicted to reach $645 billion by 2025, indicating its rapid evolution (Neiman *et al.*, 2024). The industry is welcoming **artificial intelligence** (AI) and machine learning (ML); according to 72% of business leaders, AI is essential (Neiman *et al.*, 2024). With an emphasis on technological control, involvement, education, and ethics, human-centred digital transformation is becoming increasingly important (Inversini, 2024). Ethical data management is essential for building trust, and the big data market was valued at $138.9 billion in 2023 (Salazar *et al.*, 2023). It is anticipated that the digital advertising market in China will reach $139.3 billion by 2027 (Mohamad, 2022), and according to forecasts by Kadam *et al.* (2023), the e-commerce zone in India is predicted to flourish from $188 billion in 2025 to $350 billion in 2030.

In the digital age, **human-centred design** (HCD), which emphasises user wants, goals, and behaviours across multiple domains, has become an essential methodology (Caballero *et al.*, 2014; Giacomin, 2014). To adapt to societal and technological developments, this paradigm places a strong emphasis on human centricity, cooperation, and prototyping (Calabretta, Kleinsmann, 2017). HCD concepts are used in digital marketing to create unified digital brands and modify tactics (Corrigan, Miller, 2011).

The structure of this paper is as follows: Section 2 presents a theoretical overview of the fundamental concepts related to human-centred and digital marketing, along with their intersections. Subsequently, it delves into metadata research drawn from the Web of Science (WoS) database and discusses the findings of the study, considering their **implications** for academia, practice, and policy. The paper concludes with an ample summary.

2 Theoretical Background

A growing number of sectors are embracing HCD to promote creative and practical solutions. HCD approaches are specifically utilised in the healthcare industry to build health apps targeted at underprivileged communities, foster goals-of-care conversations, and produce patient-focused innovations (Abedini *et al.*, 2021; Bartlett *et al.*, 2021; Göttgens, Oertelt-Prigione, 2021). Moreover, a human-centred approach found its path in the food and beverage industry, which takes into account both intrinsic and extrinsic factors that influence customer preferences. Internal sensory attributes, such as aroma, taste, mouthfeel, and aftertaste, take a noteworthy role, shaping the multisensory experience of the product, while external elements like product origin, packaging, branding, and price also affect consumer choices (Krstić *et al.*, 2021). Sensory marketing, which targets these various senses, proves to be an effective strategy for influencing preferences and enhancing consumer satisfaction (Krstić *et al.*, 2021).

Furthermore, HCD training greatly improves team relations and product development procedures in research settings (Norman *et al.*, 2021). In a similar vein, in-car warning systems that reduce driver weariness and attention are designed by the automotive industry using HCD concepts (Horberry *et al.*, 2021). Moreover, HCD is used in the industrial sector to address sociotechnical issues as the sector transitions to Industry

4.0 (Kadir, Broberg, 2020; Nguyen Ngoc *et al.*, 2021). In line with the aforementioned, a human-centred approach can also be found in the film industry. Nikolić *et al.* (2022, 2023) explain in their work that:

1. Genre preferences evolve, influenced by repeated exposure to specific genres.
2. Preferences for genres such as comedy, drama, and horror significantly impact consumer film choices.
3. Each individual has unique tastes and preferences when it comes to films.
4. Understanding audience preferences for particular genres is essential for film producers and distributors, guiding their decisions on production and marketing strategies.

In addition to the examples from other industries, the fashion industry demonstrates the importance of adapting to changing consumer behaviours. For managers in this sector, it is vital to understand how these behaviours have evolved and to identify which consumer preferences are influencing the market (Nikolić, Kostić-Stanković, 2022). By anticipating these shifts, companies can improve their operational processes and stay ahead of trends. One major shift has been the growing environmental awareness among consumers. More and more people are not only becoming knowledgeable about environmental issues, but they are also increasingly concerned about the impact of human actions on the planet (Vlastelica *et al.*, 2023a, 2023b). As a result, there is a noticeable shift in consumer attitudes toward sustainable products, including clothing, which has created new opportunities for companies that prioritise eco-friendly practices (Vlastelica *et al.*, 2023a, 2023b).

To better engage customers, digital marketing strategy makes use of data analytics, content marketing, and search engine optimisation (Tandava *et al.*, 2022). Gaining real-time insights into campaign performance through digital marketing is a major benefit that allows for quick modifications (Visser *et al.*, 2019). Digital marketing, in contrast to traditional marketing, makes use of Internet platforms to enable interactive, two-way contact between companies and customers (Visser *et al.*, 2019; Dašić *et al.*, 2023). According to Damnjanović *et al.* (2020), the main advantages are lower costs and the capacity to personalise advertising content. Nevertheless, as digital marketing evolves, issues such as data security and privacy in digital environments remain critical challenges (Abdulllah *et al.*, 2023).

The relevance of the confluence of digital marketing and human-centred methods in industrial contexts is growing. Interestingly, Industry 5.0 prioritises worker well-being and digital technology collaboration with a heavy emphasis on human-centricity (Alves *et al.*, 2023). Furthermore, Rovzanec *et al.* (2022) have presented AI frameworks that incorporate human-centricity, safety, and trustworthiness as crucial elements in Industry 5.0 applications. Moreover, industrial enterprises' digital marketing skills are always changing. New themes that emerge include sophisticated technology, social media interaction, digital partnerships, and communication channels (Herhausen *et al.*, 2020). Furthermore, big data analytics and cognitive computing are essential for advancing data-driven industrial marketing strategies (Lytras *et al.*, 2020). Notably, human-centred AI is becoming more and more recognised as essential to achieving sustainable development goals (Mhlanga, 2022).

3 Research Methodology

The methodology utilised in this study is hybrid bibliometric, incorporating both quantitative and qualitative approaches. The research process is underpinned by secondary data and is presented through a dual approach, including both theoretical and illustrative elements. A framework overview of the research methodology is illustrated in Fig. 1.

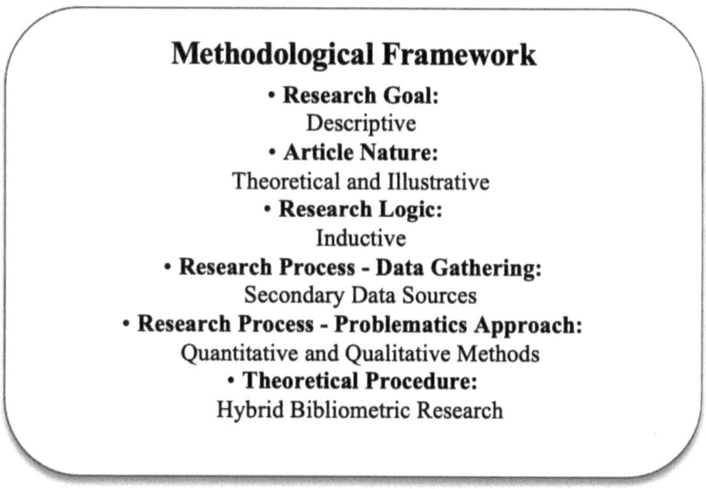

Fig. 1. Methodological framework (the authors' work adapted from Machado and Martens, 2015)

A two-stage methodological framework has been devised for a hybrid **bibliometric analysis** to extract pivotal insights from the extant scholarly corpus in this field. The preliminary phase covers a bibliometric exploration of metadata retrieved from a prominent database, furnishing a holistic overview of prevailing research directions and the intellectual landscape of the domain. Subsequently, an exacting set of criteria was employed to distil the studies unearthed through the bibliometric investigation, culminating in a refined compilation of publications apt for scrutiny. This study endeavours to conduct an exhaustive bibliometric inquiry to unearth and address lacunae within the literature, thus facilitating a foundation for prospective scholarly endeavours and in-depth exploration.

The methodological schema is initiated with the deployment of bibliometric methodologies, beginning with the extraction of data from the WoS database, renowned for its scrupulous precision and expansive coverage (Rodgers, Zhang, 2022). The search strategy incorporates exclusion parameters, concentrating on 'human-centered digital marketing (Topic)', or 'human oriented digital marketing (Topic)', or 'digital marketing and human needs (Topic)', where 'Topic' includes titles, abstracts, and keywords. The scope of selection is confined to peer-reviewed journal articles, systematic reviews, and conference proceedings composed in English, with a publication range extending from 1997 to May 6, 2024. This rigorous filtering mechanism culminates in the compilation of a corpus encompassing 542 distinct publications, which constitutes the empirical foundation for the subsequent bibliometric analysis. The analytical and interpretative processes of this dataset are executed utilising the Biblioshiny package, integrated within the R software environment. This tool facilitates a nuanced and exhaustive mapping of the bibliometric landscape in this research sphere, offering a visual and quantitative elucidation of the intellectual architecture and prevailing trends (León-Castro *et al.*, 2021; Tanwar *et al.*, 2022).

A qualitative method is incorporated into the hybrid bibliometric analysis in the second phase. To perform a systematic literature evaluation, a thorough filtering process is used for the original set of 542 publications. This study aims to derive significant insights on **human-centred digital marketing** that apply to academics, professionals in the business, and policymakers. Refined selection criteria are applied to different categories, fields, and features in the WoS database, as per the recommendations put out by Xiao and Watson (2019) (Fig. 2). The publishing pool is narrowed down to 29 articles (the datasets used and analysed during the current study are available from the corresponding author upon reasonable request) by this methodical technique, guaranteeing a focused and thorough examination of the literature.

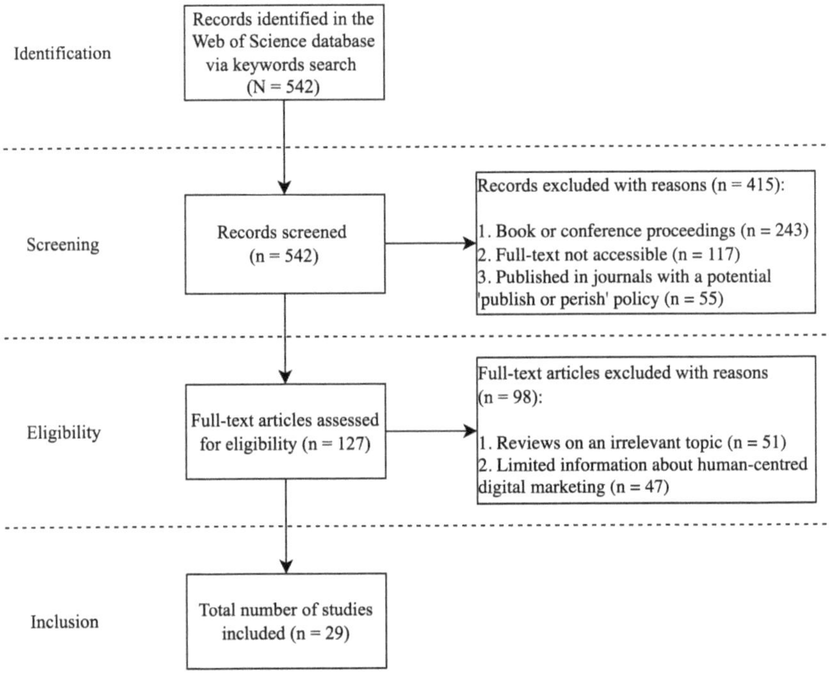

Fig. 2. Literature scrutinisation for qualitative exploration (the authors' work adapted from Xiao and Watson, 2019)

4 Research Results and Discussion

The succeeding subsections show the results and discussions of the two-fold process of examination per the methodology adopted. Initially, the bibliometric findings are presented textually and visually. The qualitative results are then synthesised, emphasising the convergence of study findings and the conclusions proposed by the authors.

4.1 Bibliometric Analysis

As outlined in Table 1, the dataset consists of 542 publications collected between 1997 and 2024. These publications originate from 457 sources. The gradual annual increase in the percentage of publications addressing human-centred digital marketing is 11.06%, and the average citation rate per paper is 11.16. A total of 1,723 authors contributed to these works, referencing 23,209 sources. The percentage of international co-authorship is 20.66%.

Table 1. The key characteristics of the articles included in the bibliometric analysis (data collected from WoS and analysed using Biblioshiny in R Studio; authors' work)

Characteristics	Results
Timespan	1997:2024
Sources	457
Documents	542
Annual Growth Rate %	11.06
Document Average Age	5.38
Average Citations per Doc	11.16
References	23,209
Keywords Plus	811
Author's Keywords	2,370
Authors	1,723
Authors of Single-authored Docs	102
Single-authored Docs	103
Co-authors per Doc	3.31
International Co-authorships %	20.66
Article	266
Article (Early Access)	17
Article (Proceedings Paper)	9
Proceedings Paper	221
Review	28
Review (Early Access)	1

Figure 3 presents the trajectory and volume of publications on human-centred digital marketing. Starting in 2012, there is a clear upward trend, with the number of publications peaking at 80 in 2021. It is crucial to note that since 2024 is still underway, the current publication count remains substantially lower.

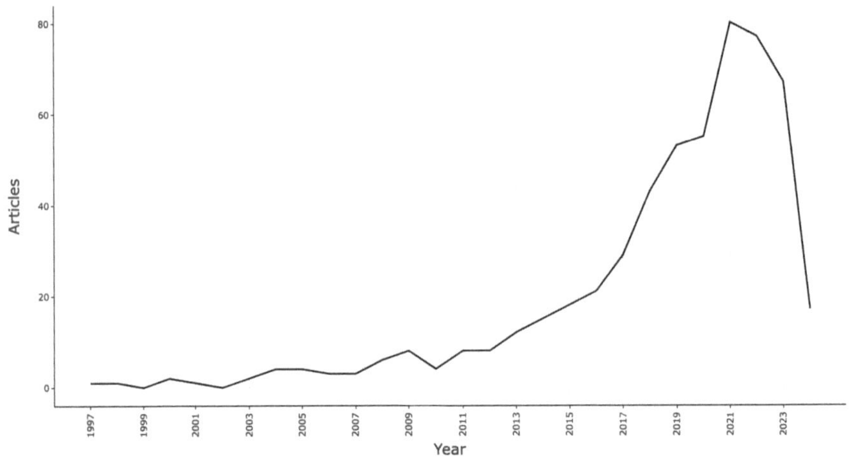

Fig. 3. Annual scientific production (data collected from WoS and analysed using Biblioshiny in R Studio; authors' work)

The following investigation, also conducted using the Biblioshiny R studio package, presents an overview of the most prominent institutions contributing to research on human-centred digital marketing. Leading the list with 15 publications from the dataset (542) is the Bucharest University of Economic Studies. Ranked second, with 12 publications, is the Polytechnic University of Madrid, Spain. In the third position, representing Eastern Europe, is Peter the Great Saint Petersburg Polytechnic University, with ten publications. The institutions are ranked in descending order by publication count, with further details on affiliations available in Table 2.

The usual scientific approach within the bibliometric analysis implies that mapping different elements of the body of knowledge, identified as a starting research point, fosters a broader understanding of the context referring to the topic scrutinised. One of the tools used for grasping a glance of mutual interdependence between those elements is a three-field plot (known as Sankey's diagram). It enables both textual and visual insight into which elements are predominant in their category and how elements from different categories are connected. For this work, three bibliometric categories are identified, which are sources of publications, keywords, and authors, respectively, from left to right (Fig. 4). The first implication drawn from Fig. 4 refers to the identification of the most occurring parties of each bibliometric category. On behalf of sources, the predominant parties are proceedings from conferences 34th Ibima and Vision 2020. On the author's side, Bannykh, G. is the author with the most occurring appearance and diverse topic exploration. On the side of keywords utilised, it appears that education and AI are the

Table 2. Most relevant affiliations (data collected from WoS and analysed using Biblioshiny in R Studio; authors' work)

Affiliations	Articles
Bucharest University of Economic Studies	15
Polytechnic University of Madrid	12
Peter the Great Saint Petersburg Polytechnic University	10
University of Aveiro	9
Delft University of Technology	8
National University of Singapore	8
Russian Presidential Academy of National Economy and Public Administration	8
University of Zagreb	8
George Mason University	7
McGill University	7

leaders, which may be a consequence of the nature and context of topics examined in previously mentioned conferences.

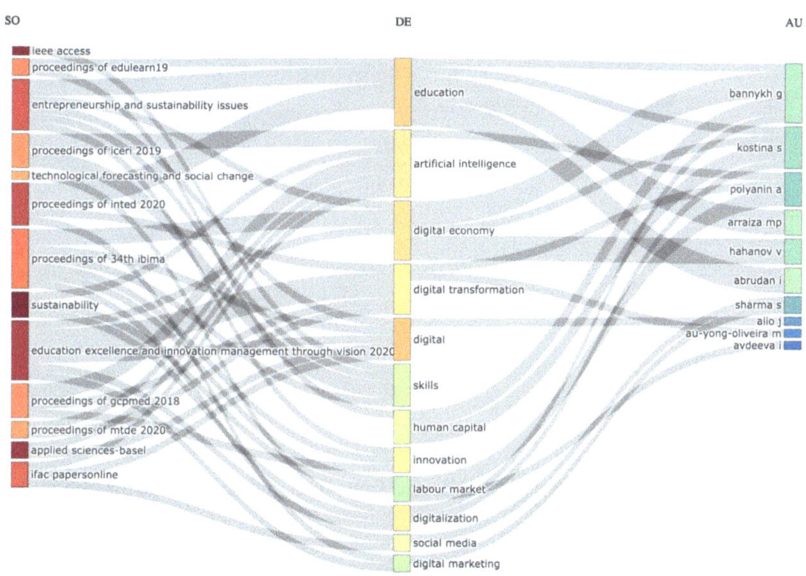

Fig. 4. Three-field plot analysis (data collected from WoS and analysed using Biblioshiny in R Studio; authors' work)

A great number of both articles and proceedings papers do not follow the expected equivalence in proportion to most occurring sources. As the three-field plot underlines, proceedings of conferences are more present and have a higher impact on this graph than articles from scientific journals. For instance, Technological Forecasting and Social Change, accompanied by Sustainability, have fewer connections to the diverse keywords compared to the sum of the proceedings from conferences such as 34th Ibima and Vision 2020. Notably, diversification between authors and their impact may be recognised as a fact that authors mainly focus on one general topic rather than having a multiversity approach, at least in this complex field of research. Particularly, authors Bannykh, G., Kostina, S., and Polyanin, A. tend to have a diverse scope when approaching the exploration of confluence between human-centred approach and digital marketing. However, other authors have only focused on specific topics. Therefore, the overall result from this figure outlines the purpose of continuous endeavour in examining the interconnection between technology-related and human-related aspects of research in the context of human-centred digital marketing.

Figure 5 presents a world map, where the parameters are defined by the existence of at least one collaborative publication between two countries. The second parameter reflects the visual representation of the connecting lines between countries, with line thickness indicating the strength of collaboration in terms of the number of publications. Based on the map, it can be concluded that the topic of human-centred digital marketing fosters global collaboration across multiple regions.

Fig. 5. Country collaboration map (data collected from WoS and analysed using Biblioshiny in R Studio; authors' work)

A deeper analysis of the collaborating countries reveals that the most frequent collaboration between countries, in terms of publications published, is between the United Kingdom and Spain, with eight papers. Then, the United States of America and India held second place with a total of seven papers. Italy and Spain, as same as the United Kingdom and Italy, each have six joint publications. Additionally, the United States of

America has six publications, each with China and Italy. Further specifics are accessible in Table 3.

Table 3. Number of publications resulting from collaboration between countries (data collected from WoS and analysed using Biblioshiny in R Studio; authors' work)

From	To	Frequency
United Kingdom	Spain	8
United States of America	India	7
Italy	Spain	6
United Kingdom	Italy	6
United States of America	China	6
United States of America	Italy	6
Spain	Netherlands	5
United Kingdom	France	5
United Kingdom	Norway	5
United States of America	United Kingdom	5

4.2 Qualitative Analysis

As detailed in the methodology section, the **qualitative analysis** was employed to pinpoint credible examples within the literature pertinent to the research topic. The depth, intricacy, and underexamined aspects of this subject are encapsulated within 29 selected publications, following a stringent sequence of selection criteria (Fig. 2). According to recent studies, different technologies and methods can significantly improve user experiences in a variety of fields. In their work, Petit *et al.* (2019) highlight how online experiences are enhanced by sensory-enabling technologies (SETs), which stimulate senses other than sight and sound. Immersion settings produced by these technologies have a good effect on customer behaviour. In keeping with a "human-centred digital marketing" (HCDM) strategy, haptic technology boosts product interaction and evaluation while visual-enabling tools bolster presence. This approach aims to make online interactions more engaging. By using a human-centred, participative approach, Patrício *et al.* (2020) investigate how service design might spur innovation in the healthcare industry. To co-create better services and imagine comprehensive, integrated healthcare systems that balance the needs of various stakeholders, this approach focuses on understanding patient experiences. An additional investigation (Wang *et al.*, 2023) explores the influence of AR on retail customer satisfaction. It makes the case that matching AR technology to consumer value orientations—like fun and social interaction—can improve contentment. This strategy caters experiences to a range of value orientations to support an HCDM viewpoint. Jin and Viswanathan (2024) have conducted studies on AI, looking at the affective aspect of its interactions and how it affects user happiness.

Table 4. Research findings and academic, practical, and policy implications of the study findings (author's analysis of the data collected from WoS)

Research findings	Implications		
	For academia	For practitioners	For policymakers
Contextual and augmented marketing exert a notable influence on marketing performance, showcasing significant impacts. Contrarily, predictive marketing demonstrates negligible effects on marketing performance. Augmented marketing emerges as the most influential factor contributing to enhanced marketing performance	Scrutinisation of utilising advanced technologies to boost marketing performance and enhance customer experiences	Augmented marketing is crucial in the Marketing 5.0 era for enhancing customer **shopping experiences** and meeting marketing objectives	Policymakers are called to facilitate the seamless integration of sophisticated, human-centred digital marketing innovations, such as augmented and contextual marketing, while upholding the sanctity of consumer data privacy and security. Equally vital is their role in encouraging expansive, cross-industry research to unravel the nuanced implications of Marketing 5.0 technologies, ensuring their adoption aligns with both ethical imperatives and transformative potential
The desire for social belonging significantly influences followers' perceptions of the personal benefits and their intentions to engage with AI-based virtual influencers	Exploring social influence on values, cultural differences in value orientations, and application of models to specific AR applications	Understanding value orientations is essential for practitioners to shape customer attitudes toward AR technology and deliver compelling customer experiences	Policymakers are tasked with ensuring AI-based virtual influencers embody inclusivity and approachability, supporting human-centred digital marketing. This requires a commitment to transparency, fairness, and ethical integrity, while addressing the broader societal impacts. AI systems should provide lasting emotional and functional support, aligning with human values, guiding regulations toward ethical and human-aligned marketing practices

(continued)

Table 4. (*continued*)

Research findings	Implications		
	For academia	For practitioners	For policymakers
Retailers should recognise how customers' value orientations shape their attitudes toward AR apps. Those open to change appreciate AR's playful and visual appeal. Usability, playfulness, and visual appeal enhance customer satisfaction		Recognising the significance of perceived usefulness, purchase process simplicity, enhanced security measured, personalised offers, and economic benefits in driving digital shopping intent and improving customer loyalty	Policymakers are urged to create flexible frameworks for the ethical use of AI in neuromarketing, prioritising privacy, consent, and autonomy. These frameworks should enable the positive potential of AI while encouraging global collaboration among regulators, industry, and academia. This synergy can help achieve fair regulations that build consumer trust and uphold corporate integrity
There exists the urge to scrutinise deeply ethical and privacy implications stemming from incorporation of AI and ML in digital marketing, marketing in general, and especially neuro-marketing as such. Furthermore, a tension between the efficacy of digital marketing techniques and the imperative of protecting consumer privacy arises, notably within the context of GDPR (general data protection regulation)			

They promote a human-centred strategy for AI-driven digital marketing, stressing the value of promoting equity and inclusivity in AI design. In line with HCDM principles, the study recommends that AI-powered virtual influencers should develop hospitable personas to improve user engagement. Silva *et al.*'s (2020) analysis of social media engagement tactics makes use of the Honeycomb Model (a framework that highlights the essential forces influencing the social media landscape, which all marketers, users, and platforms navigate) to pinpoint important elements, including reputation, presence, and identity. Although the study does not specifically address "human-centred digital marketing," it does stress the importance of comprehending and matching target audience preferences to maximise engagement. Table 4 encapsulates the research findings and articulates the proposed implications for future research directions.

5 Conclusion

In medias res, the work presented emphasises the notable urge of human-centred digital marketing in the changing digital landscape. The study used a mixed bibliometric methodology to sift through 542 articles, focusing on 29 major pieces. This highlights the growing academic interest and practical utility of human-centred approaches in digital marketing.

The results indicate that human-centred digital marketing significantly improves marketing success, with augmented marketing appearing as the most impactful component. Contextual marketing also had a significant influence, although predictive marketing had negligible results. The yearly growth rate of publications in this domain is 11.06%, with an average citation rate per document of 11.16. The degree of international collaboration is notable, with 20.66% of the papers featuring cross-border co-authorship, underlining the worldwide significance of this research.

It is imperative to recognise the constraints of this research. Selection bias may have been introduced as a result of the WoS database being the only source of relevant papers that were not included. To augment the robustness of the findings, future inquiries must incorporate a more expansive and variegated spectrum of datasets, coupled with the inclusion of a more substantial corpus of scholarly articles.

What's more, future research ought to investigate how human-centred digital marketing may use cutting-edge technologies like AR and AI. It is also important to look at how these technologies may improve and further personalise client experiences. Furthermore, a more thorough analysis of the privacy and ethical ramifications of AI and ML in digital marketing is required, especially concerning data protection regulations like GDPR.

Overall, human-centred digital marketing provides a disruptive approach that is consistent with current company structures and customer needs. With worldwide digital marketing expenditure expected to exceed $645 billion by 2025 and 72% of corporate leaders recognising AI's critical role, academics, practitioners, and governments must work together. Such partnership will ensure the creation of ethical, successful, and innovative digital marketing strategies that prioritise human-centricity while addressing the difficulties of the digital age.

References

Abdulllah, N.D.H., Tiwari, N.S., Pandey, N.D.P.: Digital marketing dynamics in the indian context: A journey through time. Int. J. Adv. Res. Sci. Commun. Techno. **1**(1), 158–161 (2023)

Abedini, N.C., et al.: Applying human-centered design to refinement of the Jumpstart guide, a clinician- and patient-facing goals-of-care discussion priming tool. J. Pain Symptom Manage. **62**(2), 1283–1288 (2021)

Alves, J., Lima, T.M., Gaspar, P.D.: Is Industry 5.0 a Human-Centred Approach? A Systematic Review. Processes. **11**(1), 193 (2023)

Bartlett, R., Boyle, J.A., Simons Smith, J., Khan, N.N., Robinson, T., Ramaswamy, R.: Evaluating human-centred design for public health: A case study on developing a healthcare app with refugee communities. Res. Involv. Engagem. **7**, 1–13 (2021)

Caballero, L., Moreno, A.M., Seffah, A.: Persona as a tool to involve humans in agile methods: Contributions from HCI and marketing. In: Seffah, A., Penzenstadler, B. (eds.) Human-Centered Software Engineering. 5th IFIP WG 13.2 International Conference, HCSE 2014, LNCS, vol. 8742, pp. 283–290. Springer, Heidelberg (2014)

Calabretta, G., Kleinsmann, M.: Technology-driven evolution of design practices: Envisioning the role of design in the digital era. J. Market. Manage. **33**(3–4), 292–304 (2017)

Corrigan, M., Miller, H.: Toward a user-centric digital ecosystem. IT Professional. **13**, 12–15 (2011)

Damnjanović, V., Lončarić, D., Dlačić, J.: Digital marketing strategy of Accor Hotels: Shaping the future of hospitality. Tour. Hosp. Manage. **26**(1), 233–244 (2020)

Dašić, D., Vučić, V., Turčinović, Ž., Tošić, M.: Digital marketing opportunities and the power of digital consumers. Ekonomika poljoprivrede. **70**(4), 1187–1199 (2023)

Giacomin, J.: What is Human Centred Design? Des. J. **17**(4), 606–623 (2014)

Göttgens, I., Oertelt-Prigione, S.: The application of human-centered design approaches in health research and innovation: A narrative review of current practices. JMIR mHealth and uHealth. **9**(12) (2021)

Herhausen, D., Miočević, D., Morgan, R.E., Kleijnen, M.H.: The digital marketing capabilities gap. Indust. Market. Manage. **90**, 276–290 (2020)

Horberry, T., et al.: Human-centered design for an in-vehicle truck driver fatigue and distraction warning system. IEEE Trans. Intell. Transport. Syst. **23**, 5350–5359 (2021)

Inversini, A.: Human centered digital transformation in travel: A Horizon 2050 paper. Tour. Rev. (2024)

Jin, S.V., Viswanathan, V.: Threatened and Empty Selves Following AI-Based Virtual Influencers: Comparison Between Followers and Non-Followers of Virtual Influencers in AI-Driven Digital Marketing. AI & SOCIETY (2024)

Kadam, S., Dhole, K., Kumar, A., Gupta, N., Mishra, S.: MIT Pune's Arts, Commerce, and Science College, Alandi, Pune, Maharashtra, India: Business Review on Amazon's Digital Marketing. Int. J. Adv. Res. Sci., Commu. Techno. **3**(1) (2023)

Kadir, B.A., Broberg, O.: Human-centered design of work systems in the transition to Industry 4.0. Appl. Ergono. **92**, 103334 (2020)

Kostić-Stanković, M., Bijakšić, S., Ćorić, N.: Influencer marketing kao način promocije brenda putem društvenih mreža. CroDiM: Int. J. Market. Sci. **3**(1), 146–158 (2020)

Krstić, J.D., Kostić, S.M.M., Veljović, S.P.: Traditional and innovative aging technologies of distilled beverages: The influence on the quality and consumer preferences of aged spirit drinks. J. Agri. Sci. (Belgrade). **66**(3), 209–230 (2021)

León-Castro, M., Rodríguez-Insuasti, H., Montalván-Burbano, N., Victor, J.A.: Bibliometrics and science mapping of digital marketing. In: Rocha, Á., Reis, J.L., Peter, M.K., Cayolla, R., Loureiro, S., Bogdanović, Z. (eds.) Marketing and Smart Technologies. Smart Innovation, Systems and Technologies, vol. 205. Springer, Singapore (2021)

Lytras, M.D., Visvizi, A., Zhang, X., Aljohani, N.R.: Cognitive computing, Big Data Analytics and data driven industrial marketing. Indust. Market. Manage. **90**, 663–666 (2020)

Machado, F.J., Martens, C.D.P.: Project management success: A bibliometric analysis. Revista de Gestão e Projetos. **06**(01), 28–44 (2015)

Mhlanga, D.: Human-centered artificial intelligence: The superlative approach to achieve sustainable development goals in the fourth industrial revolution. Sustainability. **14**(13), 7804 (2022)

Mohamad, R.: The effect of internet marketing on external and internal currency of the country. Wasit J. Compu. Mathe. Sci. **1**(3), 95–101 (2022)

Neiman, I., Dubovenko, M., Kaylyuk, O.: Current trends in digital marketing development. Eco. Soc. **63** (2024)

Nguyen Ngoc, H., Lasa, G., Iriarte, I.: Human-centred design in industry 4.0: Case study review and opportunities for future research. J. Intell. Manuf. **33**, 35–76 (2021)
Nikolić, D., Kostić-Stanković, M., Jeremić, V.: How does genre preference influence the importance of film marketing mix elements: Evidence during the COVID-19 pandemic. Eco. Res. **36**(1) (2023)
Nikolić, D., Kostić-Stanković, M., Jeremić, V.: Market segmentation in the film industry based on genre preference: The case of millennials. Eng. Eco. **33**(2), 215–228 (2022)
Nikolić, D., Kostić-Stanković, M.: Improving the economic sustainability of the fashion industry: A conceptual model proposal. Sustainability. **14**(8), 4726 (2022)
Norman, M.K., et al.: Assessing the application of human-centered design to translational research. J. Clin. Trans. Sci. **5**(1) (2021)
Patrício, L., Sangiorgi, D., Mahr, D., Čaić, M., Kalantari, S., Sundar, S.: Leveraging service design for healthcare transformation: Toward people-centered, integrated, and technology-enabled healthcare systems. J. Ser. Manage. **31**(5), 889–909 (2020)
Patricio, R., Gomide, P., Rocha, L.: Taking the digital innovation journey beyond technology: A human-centered design approach. J.Innov. Manage. **10**(4), 26–46 (2022)
Petit, O., Velasco, C., Spence, C.: Digital sensory marketing: Integrating new technologies into multisensory online experience. J. Inter. Market. **45**(1), 42–61 (2019)
Rodgers, S., Zhang, W.: Evaluating reliability of Google Scholar, Scopus, and Web of Science: A study of faculty in U.S. advertising and public relations programs. J. Mass Commu. Edu. **77**(3), 292–307 (2022)
Rovzanec, J.M., et al.: Human-centric artificial intelligence architecture for industry 5.0 applications. Int. J. Prod. Res. **61**, 6847–6872 (2022)
Salazar, D.T., Ugsha, M.G.T., Freire, A.S.C., Herrera, A.V.P.: The ethical cross-roads of personal data collection. Health, Sci. Techno.—Lect. Series. **2**, 400 (2023)
Silva, S.C., Feitosa, W., Duarte, P., Vasconcelos, M.: How to increase engagement on social media using the honeycomb model: A case study in a Portuguese HR company. Revista de Gestão. **27**(2), 153–167 (2020)
Štavljanin, V., Pantović, V.: Onlajn doživljaj kupaca—implikacije na digitalno bankarstvo [Online customer experience—implications for digital banking]. Bankarstvo. **46**(2), 100–129 (2017)
Tandava, N.A.R., Tiwadi, N.M.V., Day, J.S.: Enhancing digital marketing strategies through data analytics. J. Busi. Anal. **15**(4), 325–338 (2022)
Tanwar, A.S., Chaudhry, H., Srivastava, M.K.: Trends in influencer marketing: A review and bibliometric analysis. J. Intera. Advert. **22**(1), 1–27 (2022)
Visser, M., Sikkenga, B., Berry, M.: Digital Marketing Fundamentals: From Strategy to ROI. Taylor & Francis (2019)
Vlastelica, T., Kostić-Stanković, M., Krstić, J., Rajić, T.: Generation Z's intentions towards sustainable clothing disposal: Extending the theory of planned behavior. Polish J. Environ. Stud. **32**(3), 2345–2360 (2023a)
Vlastelica, T., Kostić-Stanković, M., Rajić, T., Krstić, J., Obradović, T.: Determinants of young adult consumers' environmentally and socially responsible apparel consumption. Sustainability. **15**(2), 1057 (2023b)
Wang, W., Cao, D., Ameen, N.: Understanding customer satisfaction of augmented reality in retail: A human value orientation and consumption value perspective. Inform. Techno. People. **36**(6), 2211–2233 (2023)
Xiao, Y., Watson, M.: Guidance on conducting a systematic literature review. J. Plann. Edu. Res. **39**(1), 93–112 (2019)

Insights into Gen-Z Values for Affective Human Resource Management

Miha Marič[1(✉)], Dejan Marinčič[2], and Anastasiya Nikolaeva Stoyanovich[1]

[1] University of Maribor, Faculty of Organizational Sciences, Kidričeva cesta 55a, 4000 Kranj, Slovenia
miha.maric@um.si

[2] Regional Development Agency of Ljubljana Urban Region, Tehnološki park 19, 1000 Ljubljana, Slovenia

Abstract. Understanding the values of employees is crucial for effective human re-source management, particularly when managing generational differences. The key changes occurred during the transition from the traditional to the modern, and then to the postmodern information society. Through generations, differences in values are perceived. Generation Z, the cohort currently entering the workforce, brings with them a distinct set of values that shape their expectations and behaviours in professional environments. This paper explores the core values of Generation Z by surveying 51 university students in Slovenia, using the Musek Values Scale to categorize their values into Apollonian and Dionysian types. The results reveal that Generation Z places the highest importance on values such as health, freedom, security, and justice, while placing less emphasis on material success and competition. These findings provide valuable insights for organizations aiming to align their HR policies with the priorities of this generation. By understanding these values, companies can better attract, engage, and retain young talent. The study concludes with recommendations for human resource management, highlighting the importance of fostering a work environment that supports personal well-being, ethical principles, and continuous learning.

Keywords: human resource management · Gen-Z · management · organization · students

1 Introduction

In today's dynamic organizational landscape, values play a pivotal role in shaping not only individual's behaviour but also the broader strategies employed in **human resource management (HRM)**. Values are an important part of every leader's daily life, as deeply held principles that guide preferences, decisions, and actions, are essential in determining how individuals perceive their work and how they engage within organizations (Jones, 2012) as well as in their personal lives. They reflect what we consider important, desirable or correct in life and work, help us understand what is important to us and set priorities in all areas of life (Zapata-Cantu, Sanguino, Barroso and Nicola-Gavrilă, 2023). There

V. Štavljanin et al. (Eds.): SymOrg 2024, LNNS 1681, pp. 93–105, 2025.
https://doi.org/10.1007/978-3-032-08090-5_8

are two known groups that form life values: Dionysian and Apollonian values (Musek, 2015). These two groups of life values are further divided into two types: Dionysian values into the hedonic type and the potency type; and Apollo values into the moral type and the fulfilment type.

As in other developed countries, value orientations in Slovenia also change with age (Marinčič et al., 2023). Each generation namely holds distinct values, shaped by the socio-economic conditions and cultural trends of their time. In the era of globalization and technological progress, understanding different generations becomes crucial for efficient human resource management. This research is important, because it builds upon the understanding of Generation Z (Gen-Z), individuals born between the late 1990s and early 2010s, a generation that is beginning to enter the workforce with unique perspectives and expectations.

Gen-Z is often characterized as "digital natives," having grown up in an era of rapid technological advancement, economic uncertainty, and heightened social consciousness. Their values reflect these experiences, differing from those of previous generations such as Millennials, Generation X, and Baby Boomers (Janssen and Carradini, 2021). For HR managers, understanding the values of Gen-Z is not only important for fostering organizational alignment but also for attracting, retaining, and motivating this emerging workforce. Research shows that values influence key HR outcomes such as job satisfaction, commitment, and leadership preferences (Scholtens and Dam, 2007; Schwartz, 2005). Therefore, aligning HRM strategies with the values of Gen-Z is essential to creating a supportive and productive work environment (Maloney et al., 2019).

Existing literature highlights generational differences in values, but there remains a gap in the specific exploration of Gen-Z's values in the context of HRM (Iorgulescu, 2016). Studies have primarily focused on their work habits, communication styles, and attitudes toward leadership. However, fewer studies have employed structured theoretical models to systematically assess the values that are most important to Gen-Z, particularly in a European context. This paper aims to fill that gap by focusing on Gen-Z's values through the lens of the Musek Values Scale, which categorizes values into Apollonian (moral and fulfilment-oriented) and Dionysian (hedonic and power-oriented) dimensions (Musek, 2015).

This study investigates the values of 51 undergraduate **students** from the University of Maribor, Slovenia, a representative sample of Gen-Z. The research explores how this generation prioritizes different values, providing valuable insights into their expectations and motivations. By identifying and analysing the key values of Gen-Z, this study aims to offer practical implications for HRM practices. Understanding these values can help organizations craft policies and work environments that resonate with this generation's priorities, ultimately fostering better employee engagement, retention, and satisfaction.

This paper contributes to the growing body of research on intergenerational differences in the workplace and offers recommendations for HR professionals seeking to adapt their management strategies to align with the values of Gen-Z. By understanding the values of this new generation, HRM can anticipate trends and proactively adjust practices to meet the evolving needs of the workforce.

2 Literature Review

Generational differences in values have been a subject of extensive research in the fields of organizational behaviour and human resource management (Marinčič et al., 2023). Traditional values in business encompass certain principles and practices that have evolved over time and are often associated with older generations (Stoyanovich and Marič, 2024). Values, as enduring beliefs that guide behaviour, influence how individuals make decisions, set priorities, and engage with their work environment (Zander et al., 2016). As each generation is shaped by the unique socio-economic, technological, and cultural contexts of their formative years, understanding these differences becomes essential for managing a multigenerational workforce (Scholtens and Dam, 2007). This review explores the importance of values in the context of human resource management (HRM) and focuses specifically on Generation Z (Gen-Z), the cohort now entering the workforce.

Values are often defined as conceptions of what is desirable or important in life, and they guide individuals' actions and decisions (Meglino and Ravlin, 1998). They also shape individuals' behaviours and their expectations of organizational practices. Meglino and Ravlin (1998) highlight that values within organizations play a significant role in dictating employees' lifestyles, work choices, and long-term goals. Furthermore, Schwartz (2005) argues that values form the basis of moral judgments, guiding individuals in understanding what is right and wrong in both personal and professional contexts. These values develop over time and are influenced by societal norms, culture, education, and personal experiences (Schwartz, 2005). Values change in individual societies over time (Baumane-Vitolina et al., 2016) and finding a unique value proposition and strategies to stand out from the crowd can be challenging (Klaczak, 2023).

The distinct values of Generation Z have garnered increasing attention in recent years. As with previous generations, Gen-Z's values reflect the environment in which they were raised. Growing up during the age of rapid technological advancement, heightened global awareness, and economic uncertainty, Gen-Z has developed a unique set of values that differentiates them from previous generations such as Millennials, Generation X, and Baby Boomers (Iorgulescu, 2016). Janssen and Carradini (2021) emphasize that Gen-Z's experiences with digital technology and social media have shaped their expectations of immediacy, transparency, and flexibility in both personal and professional settings (Ormiston and Seymour, 2011). These values are evident in their workplace preferences, where they prioritize work-life balance, autonomy, and meaningful work (Popaitoon, 2022) as they help them direct their actions and decisions in accordance with their deeper beliefs (Scholtens and Dam, 2007).

Several frameworks have been developed to study values, i.e. Rokeach Value Survey or Schwartz's Value Survey, with one of the most comprehensive models being the **Musek Values Scale** (Musek, 2015) shown in Fig. 1, which we will use. Musek (2015) classifies values into two main groups: Apollonian values, which encompass moral and fulfilment-oriented values, and Dionysian values, which include hedonic and power-oriented values. The Apollonian values reflect a focus on ethical behaviour, social responsibility, and personal growth, while the Dionysian values emphasize pleasure, material success, and status. This classification is particularly relevant for understanding Gen-Z, whose values tend to reflect a blend of moral concern for social issues and a desire for personal fulfilment (Musek, 2015).

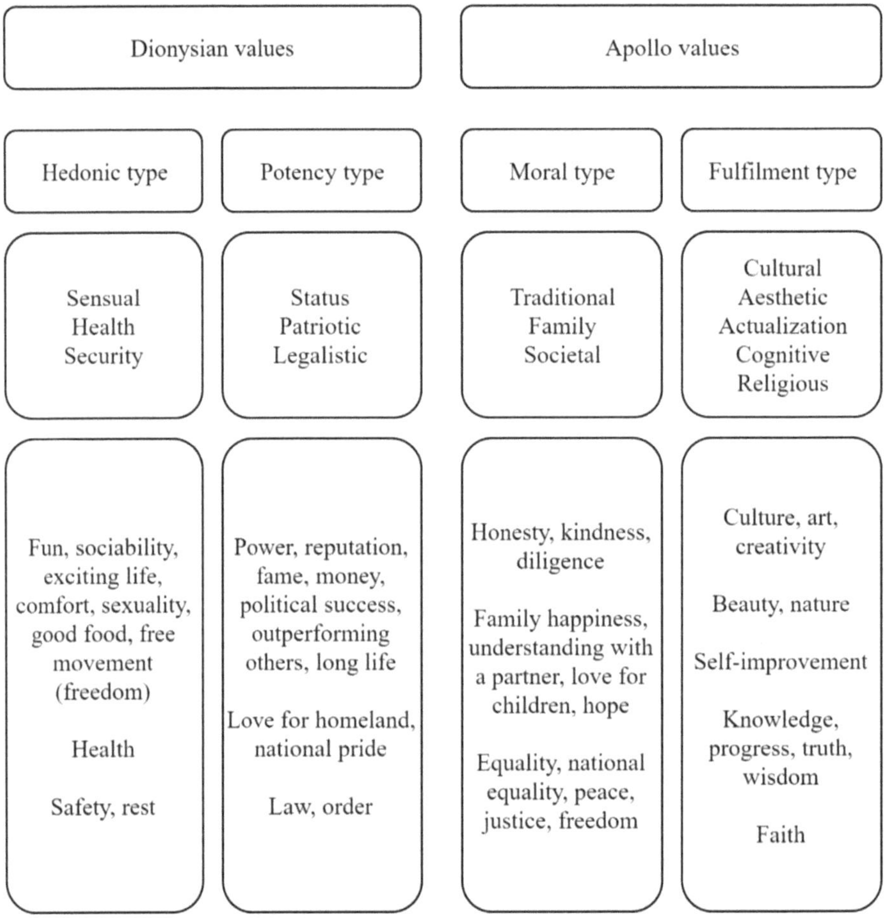

Fig. 1. Musek Values Scale (Musek, 2015).

Gen-Z's orientation toward values has been shown to differ significantly from that of previous generations. Marinčič, Đurica and Marič (2023) highlight that while older generations may prioritize traditional markers of success—such as job stability, status, and financial gain—Gen-Z places a greater emphasis on self-actualization, ethical leadership, and social justice. This shift in priorities is important for organizations to understand, as it impacts how they attract and retain talent from this generation. Values such as honesty, justice, and equality are particularly important to Gen-Z, indicating their desire to work in environments that align with their ethical principles (Scholtens and Dam, 2007).

Iorgulescu (2016) further adds that Gen-Z's value system is shaped by a complex combination of personal fulfilment and a heightened sense of social awareness. This generation is motivated by opportunities for self-improvement and professional development, as well as by the desire to contribute to social change. This presents both challenges and opportunities for HRM. For example, organizations that foster ethical practices and provide avenues for personal growth may be better positioned to engage and retain Gen-Z employees (Clegg et al., 2007).

In addition to moral values, Gen-Z's hedonic values also play a significant role in their work **preferences**. According to Musek (2015), Dionysian values such as the pursuit of happiness, comfort, and personal freedom are highly valued by this generation. Gen-Z seeks workplaces that not only provide financial security but also offer opportunities for creativity, innovation, and flexibility. Achieving alignment between decisions and values, brings confidence in actions and gives further motivation towards achieving one's goals (Clegg et al., 2007). Janssen and Carradini (2021) argue that the emphasis on these values requires organizations to rethink traditional work structures and offer more flexible, engaging work environments. The values of a community or society can influence behaviour and attitudes and can have a wider impact on social development and change (Ormiston and Seymour, 2011).

Understanding these generational shifts in values is crucial for effective HRM, as they directly influence workforce expectations and organizational dynamics. Gen-Z's emphasis on both ethical principles and personal fulfilment suggests that organizations need to create inclusive, transparent, and growth-oriented work cultures (Popaitoon, 2022). Moreover, by aligning HR strategies with the values of Gen-Z, companies can not only enhance employee engagement but also foster long-term loyalty and productivity.

In conclusion, the literature suggests that the values of Gen-Z are distinct and multifaceted, reflecting a combination of ethical, social, and hedonic orientations. Organizations that recognize and adapt to these values will be better positioned to attract, engage, and retain this new generation of talent. Further research is needed to explore how these values evolve over time and across different cultural contexts, but the current body of knowledge provides a valuable foundation for adapting HRM practices to meet the expectations of Gen-Z.

3 Methodology

Values often shape our decision-making processes (Marinčič and Marič, 2022). Therefore, by understanding one's values, we can adequately and more effectively manage them in the future. Therefore, we ask ourselves the research question: What values does the Gen-Z have? The aim of this research is to explore and analyse the core values of Generation Z (Gen-Z) and their implications for human resource management (HRM). To achieve this, a quantitative research design was employed, utilizing a survey method to collect data from Gen-Z respondents. This section outlines the research approach, sample characteristics, data collection process, and analytical methods used in the study.

A quantitative approach was chosen for this study as it allows for the systematic measurement of values and their importance among a representative sample of Gen-Z. By using structured questionnaires, the research provides a clear and objective view of the values held by Gen-Z students. The study is framed using the Musek

Values Scale (Musek, 2015), a well-established model for categorizing values into two overarching groups: Apollonian values, which are moral and fulfilment-oriented, and Dionysian values, which are hedonistic and power-oriented. This model allows for a detailed understanding of Gen-Z's value orientations and their potential influence on HR practices.

The sample for this research consists of 51 undergraduate students from the University of Maribor, Slovenia. The respondents, all of whom were born between 1999 and 2003, are representative of Gen-Z, aged 19 to 23, with the average age of the respondents being 21 years at the time of data collection. The gender distribution within the sample is almost balanced, with 25 (49.02%) male and 26 (50.98%) female participants. The students were selected from various programs within the University of Maribor's Faculty of Organizational Sciences, providing a diverse academic background, although they all share a common generational context.

Data was collected in October 2022 through an anonymous online questionnaire distributed to the students. The questionnaire was designed based on the Musek Values Scale (Musek, 2015), which categorizes values into four major types: hedonic, potency, moral, and fulfilment. The questionnaire consisted of multiple sections. The first section covered demographic information, while the subsequent sections focused on assessing respondents' Apollonian and Dionysian values using a five-point Likert scale. The scale ranged from 1 (not important at all) to 5 (absolutely important), allowing for the quantitative assessment of each value's significance to the respondents.

The Musek Values Scale (Musek, 2015), a comprehensive tool for measuring individual values, was employed in this research. The scale is organized hierarchically, with Apollonian values representing moral and fulfilment types, and Dionysian values reflecting hedonic and power-related types. This framework was selected because of its ability to capture both personal and societal values, which are particularly relevant for understanding the value orientations of Gen-Z. The Apollonian values encompass concepts such as justice, honesty, and self-actualization, while the Dionysian values include personal pleasure, power, and success.

The collected data was analysed using quantitative statistical methods. Descriptive statistics, such as mean values and standard deviations, were calculated to determine the importance of each value type to the respondents. Separate analyses were conducted for each subset of values (Apollonian and Dionysian) to allow for a more detailed understanding of the value priorities of Gen-Z students. The analysis focused on identifying the highest and lowest-rated values within each group, as well as those that occupied middle positions.

Given that the study involved human participants, ethical considerations were carefully observed throughout the research process. Participation in the study was voluntary, and the respondents were fully informed about the purpose of the research. The survey was anonymous to protect the privacy and confidentiality of the participants. All data were handled and analysed in accordance with ethical guidelines to ensure that no individual could be identified or harmed through their participation in the research.

4 Results

This section presents the findings from the survey conducted among 51 under-graduate students from the University of Maribor's Faculty of Organizational Sciences, focusing on their values as representatives of Generation Z (Gen-Z). The values were assessed using the Musek Values Scale (Musek, 2015), which categorizes values into Apollonian (moral and fulfilment-oriented) and Dionysian (hedonic and power-oriented) groups. The analysis was carried out separately for each subset, providing insights into the value orientations of Gen-Z students. The results are summarized in tables, showing the mean values and standard deviations for each set of values.

Dionysian values, which emphasize pleasure, personal power, and material success, were divided into two subgroups: hedonic and potency values. The results highlight distinct priorities within these value types.

Table 1 illustrates the respondents' ratings of the hedonic values. "Health" was the highest-rated value (mean = 4.86, SD = 0.40), indicating that Gen-Z students place a strong emphasis on physical well-being and personal care. This finding is consistent with broader trends showing increased health consciousness among younger generations. Other highly rated hedonic values included "freedom of movement" (mean = 4.67, SD = 0.55) and "security" (mean = 4.39, SD = 0.87), reflecting a desire for personal autonomy and stability. On the lower end, "exciting life" (mean = 3.37, SD = 0.94) and "sexuality" (mean = 3.47, SD = 1.14) were rated less important, suggesting that excitement and thrill-seeking are not key priorities for this cohort.

Table 1. Average values of respondents' Dionysian values—Hedonic type for Gen-Z students (n = 51).

Value	Mean	Std. Deviation
Joy	4.45	0.83
Entertainment	3.80	0.83
Sociability	4.29	0.83
Exciting life	3.37	0.94
Comfort	3.59	1.10
Sexuality	3.47	1.14
Good food	3.88	1.11
Free movement - freedom	4.67	0.55
Health	4.86	0.40
Security	4.39	0.87
Rest	4.27	0.87

Table 2 shows the average ratings for potency values. "Order" emerged as the most important value in this category (mean = 4.08, SD = 0.77), followed closely by "laws" (mean = 3.96, SD = 0.92), signalling a preference for structure, rules, and organization among Gen-Z students. On the other hand, "political success" (mean = 2.02, SD = 1.01) and "fame" (mean = 2.20, SD = 0.89) were rated the lowest, indicating that traditional markers of success, such as political influence or public recognition, hold little appeal for these respondents. This suggests a shift in values from external validation toward personal fulfilment and stability.

Table 2. Average values of respondents' Dionysian values—Potential type for Gen-Z students (n = 51).

Value	Mean	Std. Deviation
Power	3.55	0.81
Reputation	3.82	0.79
Fame	2.20	0.89
Money	3.82	0.87
Political success	2.02	1.01
Outdoing others	2.84	1.07
A long life	3.39	1.00
Love for the motherland	3.27	1.20
National pride	3.33	1.28
Order	4.08	0.77
Laws	3.96	0.92

Apollonian values, which focus on morality, self-actualization, and cultural fulfilment, were analysed in two subcategories: moral values and fulfilment values. These results reflect the ethical and intellectual priorities of Gen-Z.

Table 3 presents the results for moral values. "Justice" was the highest-rated value in this category (mean = 4.73, SD = 0.53), showing a strong commitment to fairness and ethical principles among Gen-Z students. Other important moral values included "peace" (mean = 4.71, SD = 0.64), "honesty" (mean = 4.59, SD = 0.61), and "love for children" (mean = 4.59, SD = 0.75). These results underscore the significance of ethical behaviour and harmonious relationships in both per-sonal and professional settings for this generation. Conversely, "style" was rated the lowest (mean = 4.24, SD = 0.74), suggesting that appearance or aesthetic preferences are not as crucial as deeper moral principles.

Table 4 shows the ratings for fulfilment values. "Knowledge" (mean = 4.61, SD = 0.60) and "nature" (mean = 4.61, SD = 0.63) were the most important fulfilment values, highlighting a strong appreciation for learning, personal growth, and the environment. This aligns with Gen-Z's broader concern for sustainability and continuous self-improvement.

In contrast, "faith" was rated the lowest (mean = 2.63, SD = 1.36), indicating that religious beliefs or spiritual practices are not central to the value system of these students.

Table 3. Average values of respondents' Apollonian values—Moral type for Gen-Z students (n = 51).

Value	Mean	Std. Deviation
Honesty	4.59	0.61
Goodness	4.41	0.67
Industriousness	4.27	0.78
Family happiness	4.29	1.03
Understanding with your partner	4.69	0.55
Love for children	4.59	0.75
Hope	4.45	0.78
Equality	4.45	0.76
National equality	4.59	0.67
Peace	4.71	0.64
Style	4.24	0.74
Justice	4.73	0.53

Table 4. Average values of respondents' Apollonian Values—Fulfilment type for Gen-Z students (n = 51).

Value	Mean	Std. Deviation
Culture	3.96	0.75
Art	3.94	0.86
Creativity	3.96	0.89
Beauty	3.67	0.86
Nature	4.61	0.63
Self-improvement	4.29	0.86
Knowledge	4.61	0.60
Progress	4.49	0.78
Truth	4.53	0.73
Wisdom	4.22	0.81
Faith	2.63	1.36

This finding may reflect a general trend of secularization among younger generations in many parts of the world.

The overall results suggest that Generation Z students prioritize values related to personal well-being, security, justice, and knowledge. Health, freedom, security, and justice emerged as the most highly rated values across all categories. These results indicate that

Gen-Z places a strong emphasis on ethical principles and personal fulfilment, rather than on traditional markers of success such as fame, power, or political influence.

Interestingly, values traditionally associated with material success or competition, such as "political success" and "outdoing others," were rated among the lowest, signifying a shift away from competitive, success-oriented values toward more introspective and socially responsible priorities. This could suggest a generation that is more focused on inner satisfaction, equality, and contributing to societal well-being rather than external achievements or hierarchical status.

While certain values such as health and knowledge were rated highly across the board, there was significant variability in the responses, particularly in values like "faith" and "political success," which had much lower mean scores. This suggests that within the Gen-Z cohort, there is a broad range of perspectives on values tied to traditional institutions like religion or politics. The low standard deviations in highly rated values like "health" and "justice" indicate a strong consensus among respondents about their importance, while higher deviations in values like "fame" and "political success" suggest more divided opinions.

5 Discussion

This section interprets the findings presented in the Results and examines their implications for human resource management (HRM) and organizational practices. The analysis reveals significant insights into the values of Generation Z (Gen-Z) and offers recommendations for organizations aiming to adapt to the evolving workforce landscape.

The findings indicate that Gen-Z places a high premium on values such as health, security, justice, and knowledge. These results align with broader societal trends that highlight a growing emphasis on personal well-being and ethical considerations. The highest-rated value, "health" (mean = 4.86), reflects a generation increasingly aware of the importance of physical and mental wellness. This heightened awareness may stem from the pervasive access to health information through digital platforms, underscoring the necessity for organizations to support employee wellness initiatives (Maloni et al., 2019).

In contrast, the low ratings for values traditionally associated with success, such as "political success" (mean = 2.02) and "fame" (mean = 2.20), suggest a shift in priorities. Gen-Z appears to value intrinsic motivations over extrinsic accolades, favouring stability, equality, and opportunities for meaningful contributions to society. This generational shift necessitates a revaluation of organizational practices to better align with the expectations of a workforce that prioritizes well-being and ethical behaviour (Janssen and Carradini, 2021).

Given these insights, human resource professionals should consider strategies to align organizational practices with the values of Gen-Z. Our recommendations are combined in four groups. Our first recommendation is to promote health and well-being by prioritizing employee health through wellness programs, flexible work arrangements, and resources that support physical and mental well-being. By creating an environment that values health, organizations can attract and retain Gen-Z talent (Popaitoon, 2022).

Our second recommendation is to foster ethical leadership and social responsibility by emphasizing ethical behaviour, social justice, and community involvement. Developing programs that encourage volunteerism and corporate social responsibility (CSR) initiatives can resonate with Gen-Z's strong moral values (Iorgulescu, 2016).

Third recommendation is to support learning and development through placing high value on knowledge and personal growth, as Gen-Z's seek opportunities for continuous learning. Organizations should invest in training programs, mentorship, and professional development initiatives to facilitate their career advancement and satisfy their intellectual curiosity (Zander et al., 2016). By adapting the leader's approach to each employee individually, with the central goal of recognizing and building quality mutual relationships and social exchanges (Ribič and Marič, 2023).

And fourth, by enhancing organizational culture, as a transparent and inclusive culture that emphasizes equality and respect will resonate with Gen-Z's values. HRM strategies should focus on creating a diverse and equitable workplace that reflects the values of the employees (Clegg et al., 2007). Also adapting flexible working conditions, opportunities for professional development and rewards that are in line with the values of Generation Z can help organizations attract and retain the best talent from this generation (Racolţa-Paina and Irini, 2021).

While this research provides valuable insights into Gen-Z values, it is essential to acknowledge certain limitations. The relatively small sample size of 51 students may limit the generalizability of the findings. Additionally, the focus on a single geographic location (Slovenia) means that cultural factors may influence the values identified. Future research should consider larger and more diverse samples across different regions and demographics to further validate these findings.

Further research could explore how Gen-Z's values evolve as they gain work experience and navigate different life stages. Longitudinal studies would provide a deeper understanding of whether the values identified in this study remain consistent over time or shift in response to changing societal norms and economic conditions. Additionally, comparative studies across different cultural contexts could shed light on how cultural values shape the priorities of Gen-Z in various regions.

6 Conclusion

The postmodern period is characterized by a breakdown of values or the so-called crisis of values among young people (Pandžić, 2010). The reason for this can be explained by the nature of today's youth values, which are quite fluid and changeable. However, the values of young people are a component of youth scenes and their everyday lives (Ule et al., 2000). They are not ideal but life patterns that young people win in trading with different lifestyle offers. The higher an individual has developed values and a sense of fellow human being or nature, the more ethical he is (Marinčič et al., 2023). These concepts in modern society for the individual represent the foundation of social socialization.

In conclusion, this research offers critical insights into the values held by Generation Z and their implications for human resource management. The findings suggest that Gen-Z prioritizes health, security, justice, and knowledge, reflecting a broader societal trend toward well-being and ethical considerations. As this generation begins to enter the **workforce**, **organizations** must adapt their HR practices to align with these values to attract, engage, and retain talent.

By promoting health and well-being, fostering ethical leadership, supporting continuous learning, and enhancing organizational culture, companies can create environments that resonate with Gen-Z's priorities. This alignment not only benefits employees but also contributes to a more engaged, motivated, and productive workforce (Maloni et al., 2019). Generation Z represents the future of the workforce and society, and it is crucial to adapt strategies and practices to meet their needs and expectations as well in the field of human resource management as i.e. in the field of marketing (Vesić et al., 2024).

The study's limitations, including its sample size and geographic focus, indicate the need for further research to explore these values in broader contexts and also by using other value scales than Musek's. Understanding the evolution of Gen-Z's values over time and across diverse cultural backgrounds will be essential for organizations aiming to remain competitive in a rapidly changing landscape. We also suggest differentiating rural versus urban (Keller and Owens, 2020) students regarding their values. Also, when discussing Gen-Z, we cannot forget, that internet also has an important effect on values (Hameed, 2011).

Ultimately, this research highlights the importance of recognizing and adapting to the values of Generation Z. As the future of the workforce, their priorities will shape organizational practices and influence the dynamics of work in the years to come. Embracing these values will not only help organizations attract and retain young talent but also foster a more harmonious and productive work environment for all employees.

References

Baumane-Vitolina, I., Cals, I., Sumilo, E.: Is Ethics Rational? Teleological, deontological and virtue ethics theories reconciled in the context of traditional economic decision making. Proc. Eco. Finan. **39**, 108–114 (2016)

Clegg, S., Kornberger, M., Rhodes, C.: Business ethics as practice. Br. J. Manag. **18**(2), 107–122 (2007)

Hameed, S.A.: Effect of internet drawbacks on moral and social values of users in education. Aust. J. Basic Appl. Sci. **5**(6), 372–380 (2011)

Iorgulescu, M.C.: Generation Z and its perception of work. Cross-Cult. Manage. J. **18**(01), 47–54 (2016)

Janssen, D., Carradini, S.: Generation Z workplace communication habits and expectations. IEEE Trans. Prof. Commun. **64**(2), 137–153 (2021)

Jones, V.L.: In Search of Conscious Leadership: A Qualitative Study of Postsecondary Educational Leadership Practices. San Diego State University, San Diego (2012)

Keller, E.M., Owens, G.P.: Traditional rural values and posttraumatic stress among rural and urban undergraduates. PLoS One. **15**(8), e0237578 (2020)

Klaczak, J.: Succession Planning Strategies in Polish Family-Owned Businesses (Doctoral study). Walden University, Minessota (2023)

Maloni, M., Hiatt, M.S., Campbell, S.: Understanding the work values of Gen Z business students. The Int. J. Manag. Edu. **17**(3), 100320 (2019)

Marič, M., Todorović, I., Marinčič, D.: Personal ethics perception and its relation to the expected properties of a good leader. In: Sustainable Business Management and Digital Transformation: Challenges and Opportunities in the Post-COVID Era. SymOrg 2022 Lecture notes in networks and systems, vol. 562, pp. 595–605 (2023)

Marinčič, D., Đurica, M., Marič, M.: Apollonian values among Gen-Z. IAI Academic Conference Proceedings, 18–23 (2023)

Marinčič, D., Marič, M.: Etika in vrednote študentov. In: 41st International Conference on Organizational Science Development, pp. 567–577 (2022)

Meglino, B.M., Ravlin, E.C.: Individual values in organizations: Concepts, controversies, and research. J. Manag. **24**(3), 351–389 (1998)

Musek, J.: Osebnost, vrednote in psihično blagostanje. University of Ljubljana, Faculty of Arts and Sciences (2015)

Musek, J.: Towards a comprehensive theory of values. Social Psychological Dynamics, 243–268 (2011)

Ormiston, J., Seymour, R.: Understanding value creation in social entrepreneurship: The importance of aligning mission, strategy and impact measurement. J. Soc. Entrep. **2**(2), 125–150 (2011)

Pandžić, E.: Sodobne vrednote slovenskih študentov. Ph.D. dissertation, Univerza v Ljubljani, Fakulteta za družbene vede (2010). http://dk.fdv.uni-lj.si/diplomska_dela_1/pdfs/mb11_pandzic-emina.pdf (2022) Accessed 2 April 2022

Popaitoon, P.: Fostering work meaningfulness for sustainable human resources: A study of generation Z. Sustainability. **14**(6), 3626 (2022)

Racolţa-Paina, N.D., Irini, R.D.: Generation Z in the workplace through the lenses of human resource professionals–A qualitative study. CAL. **22**(183), 78–85 (2021)

Ribič, T., Marič, M.: Leader's possession of linguistic intelligence in relation to Leader–Member Exchange theory. J. Intelligence, 11(5), 92 (2023)

Scholtens, B., Dam, L.: Cultural values and international differences in business ethics. J. Bus. Ethics. **75**, 273–284 (2007)

Schwartz, M.S.: Universal moral values for corporate codes of ethics. J. Bus. Ethics. **59**, 27–44 (2005)

Stoyanovich, A.N., Marič, M.: Traditional values in the digital era: case study of a family business. In: Conference Proceedings of the 43rd International Conference on Organizational Science Development: Green and Digital Transition—Challenge or Opportunity, pp. 695–707. University of Maribor, University Press, Maribor (2024)

Ule, M., Rener, T., Mencin Čeplak, M., Tivadar, B.: O. Socialna ranljivost mladih. Ministrstvo za šolstvo in šport. Urad Republike Slovenije za mladino. (2000)

Vesić, G., Arsenijević, G., Marič, M.: Digital Political Marketing. Pearson, Cop., Harlow (2024)

Zapata-Cantu, L., Sanguino, R., Barroso, A., Nicola-Gavrilă, L.: Family business adapting a new digital-based economy: Opportunities and challenges for future research. J. Knowl. Econ. **14**(1), 408–425 (2022)

Zander, L., Jonsen, K., Mockaitis, A.I.: Leveraging values in global organizations: Premises, paradoxes and progress. Manag. Int. Rev. **56**(2), 149–169 (2016)

Gen Z Career Choices: A Cross-Country Analysis of Key Drivers

Miha Marič[1], Slavica Cicvarić Kostić[2](✉), and Ivan Todorović[2]

[1] Faculty of Organizational Sciences, University of Maribor, Kidričeva cesta 55a, 4000 Kranj, Slovenia

[2] Faculty of Organizational Sciences, University of Belgrade, Jove Ilića 154, 11000 Belgradc, Serbia

slavica.cicvaric.kostic@fon.bg.ac.rs

Abstract. With Generation Z entering the global workforce, it has become increasingly important for companies to understand the traits, career aspirations, and expectations that drive their career choice. This paper explores the main drivers of career choice among Gen Z students, which are related to the elements and values they expect regarding jobs. The research included 119 students enrolled in undergraduate or master's programs at two universities in the same field of study but situated in distinct Central and Eastern European countries, the Republic of Slovenia and the Republic of Serbia. Data collection took place in April–May 2024. The results show that the main drivers for career choice in Slovenia and Serbia are Life Balance, Freedom, and Expertise, but in different order. At the same time, Organization Membership and Entrepreneurship are the lowest-ranked values in Slovenia, and Entrepreneurship and Management are in Serbia. The paper further examines the influence of national cultures on the identified career expectation themes among students in Slovenia and Serbia. By exploring Gen Z's expectations regarding careers, the research presented in the paper could give valuable insights for businesses to shape their strategies to meet this generation's evolving needs and aspirations, ensuring pathways to meaningful and sustainable careers.

Keywords: career aspirations · career expectations · career choices · national culture · students · wage expectations · salary · Gen Z · Slovenia · Serbia

1 Introduction

A generation is a group of individuals born in a certain time environment and belonging to the same age group (Kovačević and Marič, 2024). Generation Z (**Gen Z**), born between the mid-1990s and early 2010s, exhibits distinct aspirations and expectations regarding the business environment (Barhate and Dirani, 2022). Career aspirations (what an individual desires to achieve in their career) and expectations (what an individual expects to happen given the current context) are essential in shaping career development and decision-making regarding career choices (Gottfredson, 2005). As Gen Zers enter higher education and the workforce, understanding their career aspirations and

V. Štavljanin et al. (Eds.): SymOrg 2024, LNNS 1681, pp. 106–122, 2025.
https://doi.org/10.1007/978-3-032-08090-5_9

expectations becomes highly relevant to employers and could give valuable insights for businesses to shape and communicate their strategies to meet this generation's evolving needs and aspirations (Cicvarić Kostić and Gavrilović Šarenac, 2019). The investigation of students' career preferences and the underlying determinants is consistently of interest to research within human resources (Milinković et al., 2017).

Grounding on the Expectancy theory (Vroom, 1964) and the Social Cognitive Career Theory (Lent et al., 1994), this paper explores career expectations among Gen Z students in two countries, highlighting key elements influencing their perceptions and attitudes. Even though there is research on this topic, there is still room for further explanation and understanding of cross-national Gen Z students' key drivers for career choices.

The research focuses on understanding the elements and values most important to Gen Z student's careers and should, therefore, guide their career choices. The data required for the analysis are gathered through an online survey among students enrolled in undergraduate or master's programs at two universities in the same field of study but situated in distinct Central and Eastern European countries, the Republic of Slovenia and the Republic of Serbia. By understanding Gen Z students' career expectations, employers can tailor job opportunities to align with their interests and goals.

2 Theoretical Background

The career landscape has become increasingly complex due to various factors. Key drivers of change, among others, include globalization, technological advancements, demographic shifts, changing work values, particularly among younger generations (Millennials and Gen Z), economic instability and uncertainty, as well as the rise of gig economy and nontraditional work arrangements (Friedman, 2005; Twenge, 2010; Brynjolfsson and McAfee, 2014; Buchanan and Yu, 2018; Abbey, 2021). These factors make career planning less predictable and demand that individuals continuously develop new skills and adapt to changing conditions.

Career aspirations and expectations are vital to career development, primarily influencing career choices (Gottfredson, 2005). Both idealistic aspirations and realistic expectations form critical parts of vocational identity and influence career-related decisions (Vroom, 1964; Lent et al., 1994; Savickas, 2002 Barhate and Dirani, 2022). According to Mayrhofer et al. (2005), career aspirations represent the intensity of an individual's commitment to participating within a specific career domain. These aspirations encompass needs, motivations, and behavioural intentions that individuals express concerning their engagement with various career fields. In addition, they refer to the long-term career goals, ambitions, and desires that individuals aim to achieve throughout their professional lives. They are shaped by personal interests, social influences, personality traits, and environmental factors such as opportunities in the labour market or socio-economic background (Kidd, 2008). While aspirations reflect an individual's long-term desires for their professional lives, career expectations are grounded in what an individual expects to happen and what is realistically attainable given the current circumstances (Saha and Sikora, 2008). Thus, expectations provide a realistic framework for achieving those goals within external conditions.

Expectancy theory (Vroom, 1964) is a key theoretical framework for understanding career aspirations and expectations. It suggests that an individual's motivation to engage in a particular behaviour is determined by their belief that the behaviour will lead to a desired outcome. This theory provides valuable insights into how individuals form and act upon their expectations in various contexts, including the business environment. Another theoretical framework is the Social Cognitive Career Theory (Lent et al., 1994), which emphasizes the role of self-efficacy beliefs, outcome expectations, and personal goals in shaping career-related decisions and aspirations. According to this theory, individuals with higher self-efficacy (belief in their ability to perform well in a career) and positive outcome expectations are more likely to set ambitious job aspirations and take action to achieve them.

Generational differences significantly impact career aspirations and expectations. Members of Generation Z, or as they are also called, iGeneration, Gen Tech, Online Generation, Post Millennials, "always clicking", "digital natives" or "mobile natives" (Dolot, 2018), were born between the mid-1990s and early 2010s when the most significant changes in the world began to unfold with the web, the internet, smartphones, laptops, freely accessible networks, and digital media (Singh and Dangmei, 2016). They have been noted for their digital fluency and preference for flexibility, purpose-driven work, as well as work-life balance, which is recognized as one of the main issues in the contemporary working environment (Marič et al., 2021). Their job aspirations are often influenced by rapid technological change and globalization, with a focus on innovation and remote work flexibility (Twenge, 2017; Seemiller and Grace, 2018). Unlike older generations, Gen Z prioritizes careers that align with their values, such as sustainability and social justice, alongside traditional aspirations like salary and security (Barhate and Dirani, 2022). Technology has a great impact on their lives, which is also reflected in the way they communicate with colleagues and customers, which means that this takes place primarily through digital communication channels (Vesić, Arsenijević and Marič, 2024).

Gen Zers have been entering the workforce, so exploring their expectations regarding the business environment is relevant, particularly elements and values relevant to their career choices. These expectations are reshaping the business landscape, prompting businesses to meet the evolving demands of this generation. Understanding the career expectations of Gen Z students requires consideration of various factors and work values, which present what is essential for them in their working lives. As working values are regarded as one of the central constructs in career decision-making, Kuron et al. (2015) grouped them into four categories. The first group refers to psychological satisfaction with work - intrinsic work values (such as interesting work, intellectual stimulation, and others); the second group is related to the material aspect of work - extrinsic work values (such as salary, job security, and others); the third group comprises social work values (such as relations with coworkers, contribution to society and other); and the fourth group refers to status values, influence, and power.

Gen Zers, having grown up in the digital age, are well-informed and interconnected, valuing transparency and accountability from businesses (Seemiller and Grace, 2018). Consequently, they have work values that are typically different for each generation. The ubiquity of technology has shaped Gen Z's career expectations, with a preference for tech-oriented fields (Twenge, 2017). As digital natives, they value jobs that offer opportunities for creativity, innovation, and remote work flexibility (Alferjany and Alias, 2020;

Janin, 2022). Even though Gen Zers value job security and stability (Gabrielova and Buchko, 2021), they often consider other relevant factors, such as flexibility, continuous learning, purpose, and adaptability in their career decisions.

They are reshaping the traditional notions of job security and are more focused on building dynamic and fulfilling careers that align with their values and aspirations. In addition, some researchers find that they easily change jobs, seeking a job that will provide them with a sense of purpose (Seemiller and Grace, 2018; Fratričová and Kirchmayer, 2018). Gen Zers are motivated by a desire to make a positive impact on society, influencing their career decisions (Tkalac Verčič and Verčič, 2024; Gaidhani et al., 2019). They seek opportunities for meaningful work aligned with their values, such as sustainability, diversity, and social justice, and they prioritize authenticity, ethical practices, and social responsibility in the business environment (Nguyen Ngoc et al., 2022; Deloitte, 2020, Jenkins, 2019).

Regarding extrinsic work values, particularly salary, they have diverse expectations influenced by various factors such as technological advancements, economic conditions, individual characteristics, social values, and others (Fernandes et al., 2021; Silva and Carvalho, 2021). They may have higher salary expectations, and many prioritize earning potential and financial security (Handshake, 2020) but face the reality of entry-level wages and competitive job markets (Twenge, 2017). On the other side, Gen Z considers self-employment as a way of professional action, mainly because they consider it better paid and gives a sense of independence (Pauli et al., 2019; Maurer, 2016). Some have entrepreneurial intentions (Todorović et al., 2024), even though it has already been shown that entrepreneurial ventures, particularly those tech-oriented, are risky (Okanović et al., 2023).

These expectations are reshaping the business landscape, and it has become pivotal for organizations to comprehend this emerging generation's characteristics and aspirations to fulfill their needs as employees (Barhate and Dirani, 2022). Our paper focuses on the elements and values significant to Gen Z students' career choices, whose understanding is essential for attracting and retaining talents in the contemporary workplace.

As cultural and economic contexts also shape career aspirations and expectations (Hofstede, 1980; Blustein, 2006), this paper explores the elements and values most important to Gen Z student's careers in two different cultural contexts: the Republic of Slovenia and the Republic of Serbia. The countries selected for the study were formerly part of the Socialist Federal Republic of Yugoslavia (SFRY), yet they now exhibit significant socioeconomic and cultural differences. These variations include disparities in macroeconomic indicators, labour markets, education systems (Rakić et al. 2019), and cultural differences (Nedeljković et al. 2018). The transition from traditional to contemporary society changed the perception and understanding of various business concepts (Marič et al., 2023), and these countries have different paths in the transition process. Accordingly, they represent an exciting sample for comparative research in students' career choice drivers.

3 Methodology

This paper comprises the analysis of relevant theoretical frameworks and the results of a survey. The survey was conducted in April–May 2024 using an anonymous questionnaire that included three sets of questions. The first was aimed at collecting data on the demographic features of the respondents, the second was aimed at collecting data on career expectations, while the third set examined **wage** expectations, as one additional aspect of career expectation that shows how students perceive their anticipated salary on the beginning and at certain point of their careers.

The Career Expectations Questionnaire (career-expectations.pdf, 2024) is compiled from 24 statements, which were measured on a four-point Likert-type scale. The statements relate to eight themes (competition, freedom, management, life balance, organization membership, expertise, learning, and entrepreneurship) that refer to the elements and values of work essential for choosing an appropriate career. The total score for each theme was obtained by summing up all three scores related to the theme, with the highest ratings indicating what is most important to one's career and, therefore, should guide their career-making decisions. The questionnaire also included questions on expected wages after students complete their studies and their wage expectations ten years after the completion of studies. We will compare their answers with each country's official average and median wage statistics to see if there are any overly high expectations.

The sample we analysed encompasses 119 students enrolled in undergraduate or master's programs at two universities in the same field of study but situated in distinct Central and Eastern European countries. Of those students, 37 (31.1%) were from Slovenia and 82 (62.9%) were from Serbia. All respondents are affiliated with faculties of organizational sciences, focusing on informatics and management. The sample included 109 undergraduate (91.6%) and ten master (8.4%) students. Totally, 43 male (36.1%) and 76 female (63.9%) students participated in the survey. Of that, 22 male respondents come from Slovenia and 21 from Serbia, while 15 female students are from Slovenia and 61 from Serbia. The average age of the respondents was 23 years (22.6 Slovenia, 23.2 Serbia), which means that they are representatives of Gen Z – their years of birth ranged from 1997 to 2004. The work experience of the students who participated in the survey is presented in Table 1.

Table 1. Work experience of participants (n = 109).

Country	Slovenia		Serbia		Total sample	
Work experience	n	%	n	%	n	%
part-time	4	10.80%	26	31.70%	30	25.21%
full-time	1	2.70%	10	12.20%	11	9.24%
freelance	2	5.40%	5	6.10%	7	5.88%
work through student services	29	78.40%	27	32.90%	56	47.06%
never worked or are volunteering	1	2.70%	14	17.10%	15	12.61%

We used statistical methods such as average value (mean), standard deviations, and correlation analysis to explore career and wage expectations by analyzing the gathered data. The results are shown in the following section.

4 Results

In the following tables, we will present the results of the average value of the respondents' agreement with the statements regarding career expectations and the average value of the respondents' wage expectations. Table 2 and 3 show the average value of the respondents' agreement with statements regarding their career expectations by country - Slovenia and Serbia.

Table 2. Average values of respondents' career expectations for Slovenia (n1 = 37).

Career expectation	Valid	Mean	Std. Deviation
1. Promotion	37	3.1081	0.87508
2. Control over how and when I work	37	3.432	0.83017
3. Being able to get a job done well through managing the efforts of others	36	2.9444	0.71492
4. Enough leisure time to travel, relax and be myself	36	3.3611	0,83333
5. Being able to contribute new ideas, which will help me build the future	37	3.2432	0.72286
6. A balance between work and other areas of my life	36	3.5833	0,76997
7. Leading a team on key organizational projects	37	3.1892	0.81096
8. Opening up new business directions through initiating new ideas	37	3.0000	0.91287
9. Being part of an organization	37	2.7027	0.93882
10. Being given challenges, which stretch me intellectually	37	3.1622	0.79977
11. Being able to show, that I have more to offer than my colleagues	37	2.9189	0.82927
12. Being able to identify closely with an organization	36	2.5278	0.87786
13. To be recognized for my expertise	37	3.1622	0.76425
14. Being able to get the most out of people in order to achieve the set goal	37	2,8378	0.89795
15. Taking the risk of getting a new business venture off the ground	37	2,7297	0.93240
16. Being able to put work in its place as an important, but not the only part of my life	37	3.1351	0.82199
17. To have the status, that comes with being part of a successful company	37	2.7568	0.86299

(continued)

Table 2. (*continued*)

Career expectation	Valid	Mean	Std. Deviation
18. To be involved in assignments, which will take the organization forward	36	3.0833	0.76997
19. To be able to see, that I am doing better, than those I am in competition with	36	2.9444	0.95452
20. Knowing, that I am respected for the specialist skills that I bring	36	3.3333	0.75593
21. Being able to work, when and where I want, so long as I can deliver results	36	3.3611	0.89929
22. Knowing every year, that I have further developed my expertise	37	3.4054	0.76229
23. Being able to make decisions without being controlled by organizational bureaucracy	37	3.3243	0.81833
24. The excitement of creating something new, whose success depends on me	37	3.0270	0.86559

Table 3. Average values of respondents' career expectations for Serbia (n2 = 82).

Career expectation	Valid	Mean	Std. Deviation
1. Promotion	81	3.3951	0.70141
2. Control over how and when I work	81	3.3704	0.67905
3. Being able to get a job done well through managing the efforts of others	81	2.9506	0.68741
4. Enough leisure time to travel, relax and be myself	81	3.7037	0,53489
5. Being able to contribute new ideas, which will help me build the future	82	3.4756	0.68899
6. A balance between work and other areas of my life	81	3.7778	0.47434
7. Leading a team on key organizational projects	81	2.8148	0.92346
8. Opening up new business directions through initiating new ideas	82	2.8902	0.75369
9. Being part of an organization	82	3.2073	0.89908
10. Being given challenges, which stretch me intellectually	81	3.3210	0.72158
11. Being able to show, that I have more to offer than my colleagues	82	2.7805	0.84655
12. Being able to identify closely with an organization	81	2.8519	0.90982

(*continued*)

Table 3. (*continued*)

Career expectation	Valid	Mean	Std. Deviation
13. To be recognized for my expertise	81	3.4815	0.69121
14. Being able to get the most out of people in order to achieve the set goal	81	3.0247	0.77420
15. Taking the risk of getting a new business venture off the ground	82	2.3659	0.82420
16. Being able to put work in its place as an important, but not the only part of my life	81	3.4444	0.67082
17. To have the status, that comes with being part of a successful company	82	2.9756	0.90234
18. To be involved in assignments, which will take the organization forward	82	3.1463	0.77175
19. To be able to see, that I am doing better, than those I am in competition with	82	2.9390	0.93423
20. Knowing, that I am respected for the specialist skills that I bring	82	3.098	0.69843
21. Being able to work, when and where I want, so long as I can deliver results	81	3.4321	0.72350
22. Knowing every year, that I have further developed my expertise	82	3.4878	0.68932
23. Being able to make decisions without being controlled by organizational bureaucracy	82	3.2561	0.76676
24. The excitement of creating something new, whose success depends on me	82	3.0610	0.72602

Tables 4 and 5 present eight different themes, where the highest ratings indicate elements and values most important to one's career and should, therefore, guide their career-making decisions by country. Themes were calculated by summing up statements included in each theme from Tables 2 and 3, as noted in the brackets by each theme in Tables 4 and 5. We can see that the themes with the highest ratings in Slovenia are Life Balance, Freedom, and Expertise, whereas Organization Membership and Entrepreneurship were the lowest-rated themes. On the other hand, in Serbia, the students value the same items, but in a different order; Life Balance is the top-ranked. Still, Expertise is before Freedom, while Entrepreneurship is with lowest ranking, and the next one is Management.

Tables 6 and 7 present the wage expectations after completing the studies and starting to work and the wage expectations (by country), when 10 years after they complete their studies. Based on our survey, the expected wage after completing the studies and starting to work among Slovenian students is around 1,800 EUR and the expected wage, when they will have 10 years of work experience is around 4,000 EUR. Serbian students expect a significantly lower starting wage, less than 1,000 EUR, but after 10 years they expect

Table 4. Average values by career expectation themes for Gen Z students - Slovenia (n1 = 37).

Themes (statements included in each theme)	Valid	Mean	Std. Deviation
Competition (1 + 11 + 19)	36	8.9444	1.73937
Freedom (2 + 21 + 23)	36	9.8889	2.14846
Management (3 + 7 + 14)	36	8.9722	1.85913
Life Balance (4 + 6 + 16)	36	10.1389	1.79129
Organization Membership (9 + 12 + 17)	36	8.0278	2.26130
Expertise (13 + 20 + 22)	36	9.8611	1.79129
Learning (5 + 10 + 18)	36	9.4722	1.79660
Entrepreneurship (8 + 15 + 24)	37	8.7568	2.20360

Table 5. Average values by career expectation themes for Gen Z students - Serbia (n2 = 82).

Themes (statements included in each theme)	Valid	Mean	Std. Deviation
Competition (1 + 11 + 19)	81	9.1358	1.87585
Freedom (2 + 21 + 23)	80	10.0500	1.51699
Management (3 + 7 + 14)	80	8.8125	1.74366
Life Balance (4 + 6 + 16)	81	10.9259	1.20185
Organization Membership (9 + 12 + 17)	81	9.0123	2.08862
Expertise (13 + 20 + 22)	81	10.5926	1.53930
Learning (5 + 10 + 18)	81	9.9506	1.70221
Entrepreneurship (8 + 15 + 24)	82	8.3171	1.75596

it to grow 3.5 times to around 3400 EUR, which is a much larger gap than in Slovenia where this ratio is around 2.2 times regarding salary increase in 10 years.

Table 6. Wage expectations - Slovenia (n1 = 37).

Themes	Expected wages		
	Valid	Mean	Std. Dev.
My wage expectations after I complete my studies and start working is (in EUR)	36	1811.1111	460.29666
My wage expectations, when I will be with 10 years of work experience - or 10 years after I complete my studies is (in EUR)	36	4079.0833	2482.69066

Table 7. Wage expectations - Serbia (n2 = 82).

Themes	Expected wages		
	Valid	Mean	Std. Dev.
My wage expectations after I complete my studies and start working is (in EUR)	82	970.6098	303.21873
My wage expectations, when I will be with 10 years of work experience - or 10 years after I complete my studies is (in EUR)	82	3395.1220	1647.21257

We have already combined all the statements representing each theme in Tables 4 and 5 and then performed a correlation analysis between career expectations and expected wages, as presented in Tables 8 and 9. We can notice that in both countries, there is a positive and statistically significant relation between wage expectations after completing studies and starting to work and wage expectations when they have 10 years of work experience. We can also notice from the correlation analysis which of the themes have a positive relation to the expected wages and which have a negative relation.

Table 8. Pearson correlation between career expectation themes and expected wages - Slovenia (n1 = 37).

		My wage expectations after I complete my studies and start working is (in EUR)	My wage expectations, when I will be with 10 years of work experience - or 10 years after I complete my studies is (in EUR)
My wage expectations, when I will be with 10 years of work experience - or 10 years after I complete my studies is (in EUR)	Pearson Correlation	.471[a]	1
	Sig. (2-tailed)	0.004	
Competition	Pearson Correlation	0.184	.342[b]
	Sig. (2-tailed)	0.290	0.044
Freedom	Pearson Correlation	0.053	0.018
	Sig. (2-tailed)	0.761	0.917
Management	Pearson Correlation	0.242	0.301
	Sig. (2-tailed)	0.161	0.079

(continued)

Table 8. (*continued*)

		My wage expectations after I complete my studies and start working is (in EUR)	My wage expectations, when I will be with 10 years of work experience - or 10 years after I complete my studies is (in EUR)
Life Balance	Pearson Correlation	−0.078	−0.281
	Sig. (2-tailed)	0.656	0.102
Organization Membership	Pearson Correlation	0.197	0.028
	Sig. (2-tailed)	0.257	0.873
Expertise	Pearson Correlation	0.034	0.273
	Sig. (2-tailed)	0.847	0.112
Learning	Pearson Correlation	0.019	0.189
	Sig. (2-tailed)	0.913	0.276
Entrepreneurship	Pearson Correlation	0.173	.420[b]
	Sig. (2-tailed)	0.314	0.011

a. *Correlation is significant at the 0.01 level (2-tailed).*
b. *Correlation is significant at the 0.05 level (2-tailed).*

By analyzing the collected data based on students' expectations, we have determined their career and wage expectations. We have also compared their wage expectations to each county's average and median wages. The results show that in Slovenia, there are positive and statistically significant relations between expectation themes Competition and Entrepreneurship and wage expectations after ten years of work, and in Serbia, between Entrepreneurship and initial wage expectations and Organization Membership and wage expectations after ten years of work.

5 Discussion and Conclusion

The results showed differences in students' career expectations between the two countries. We can see that the themes with the highest ratings are Life Balance, followed by Expertise and Freedom in Slovenia, and Freedom and Expertise in Serbia. In contrast, the lowest-rated themes were Organization Membership and Entrepreneurship in Slovenia and Entrepreneurship and Management in Serbia, and thus, we provided the answer to the first research question. Both distinctions and similarities presented in Table 10 can be explained through cultural backgrounds. **National culture** overviews (Geert Hofstede Country comparison graphs) of the two countries are shown in Fig. 1.

Table 9. Pearson correlation between career expectation themes and expected wages - Serbia (n2 = 82).

		My wage expectations after I complete my studies and start working is (in EUR)	My wage expectations, when I will be with 10 years of work experience - or 10 years after I complete my studies is (in EUR)
My wage expectations, when I will be with 10 years of work experience - or 10 years after I complete my studies is (in EUR)	Pearson Correlation	.367[a]	1
	Sig. (2-tailed)	0.001	
Competition	Pearson Correlation	0.163	0.114
	Sig. (2-tailed)	0.145	0.310
Freedom	Pearson Correlation	−0.007	−0.008
	Sig. (2-tailed)	0,950	0,945
Management	Pearson Correlation	0,091	0,108
	Sig. (2-tailed)	0.425	0.341
Life Balance	Pearson Correlation	−0.033	0.103
	Sig. (2-tailed)	0.771	0.359
Organization Membership	Pearson Correlation	0.080	,248[b]
	Sig. (2-tailed)	0.476	0,025
Expertise	Pearson Correlation	-.,012	0.092
	Sig. (2-tailed)	0.918	0.416
Learning	Pearson Correlation	0.074	0.117
	Sig. (2-tailed)	0,511	0.299
Entrepreneurship	Pearson Correlation	.237[b]	0.140
	Sig. (2-tailed)	0.032	0.211

a. *Correlation is significant at the 0.01 level (2-tailed).*
b. *Correlation is significant at the 0.05 level (2-tailed).*

As uncertainty avoidance is highly expressed in both countries, it explains why Entrepreneurship theme is among the lowest rated. Slovenian people are more long-term oriented, which may be the factor of the more expressed theme of Management among Slovenian students compared to those from Serbia. On the other hand, the results show higher scores for Competition, Learning, and Expertise in Serbia, which can be clarified through increased masculinity at the national level, which reflects a larger motivation

Table 10. Comparative analysis of career expectations among students from Slovenia and Serbia.

	SLOVENIA	SERBIA
Competition (1+11+19)	8,9444	9,1358
Freedom (2+21+23)	9,8889	10,05
Management (3+7+14)	8,9722	8,8125
Life Balance (4+6+16)	10,1389	10,9259
Organization Membership (9+12+17)	8,0278	9,0123
Expertise (13+20+22)	9,8611	10,5926
Learning (5+10+18)	9,4722	9,9506
Entrepreneurship (8+15+24)	8,7568	8,3171

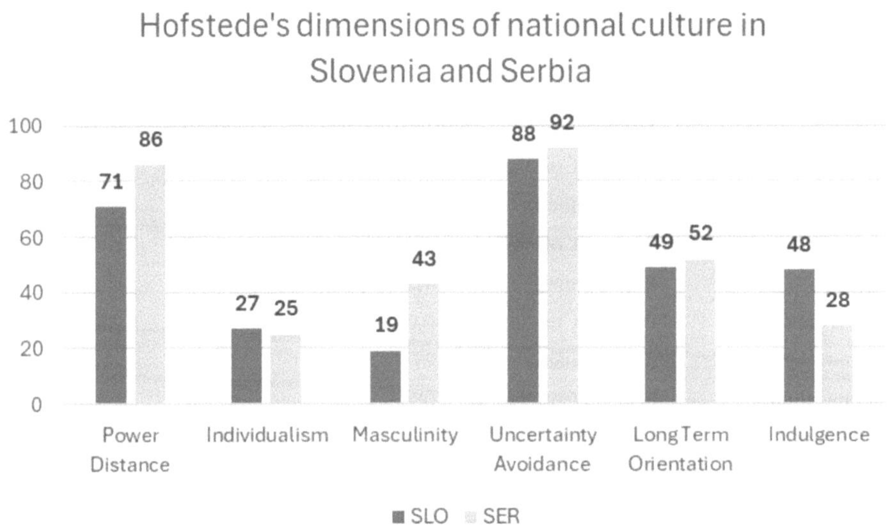

Fig. 1. National culture comparison between Slovenia and Serbia (authors' illustration based on the results from Geert Hofstede Country comparison graphs).

towards achievement and success. Variety regarding Life Balance might be clarified through different indulgence at the national level. Finally, Freedom, as the theme that was among the top rated in both countries, is most probably caused by the specifics of Gen Z, as cultural background might implicate the opposite result, having in mind the power distance dimension in national cultures. As explained in the theoretical part of this paper, new generations have different mindsets and expectations than previous ones. Their motivation is primarily to preserve their lifestyle and free time; their career goals do not focus on just one but aim for multiple careers simultaneously (Kovačević and Marič, 2024). The findings from the study emphasize how Gen Z's career aspirations are shaped by their local cultural contexts. Multinational companies and other businesses operating in these two countries should consider how these differences manifest and adapt their

HR practices accordingly. In addition, based on the study's results, companies can craft their employer branding messages to align with local cultural priorities.

Each theme positively or negatively affects the expected wage in both countries, which answers the second research question. However, not all relations are statistically significant. Differences in expected wages after studies can be explained by different levels of economic development between the two countries. When comparing their wage expectations, we included the official statistical data for the average wages in each country, which were for February 2024, 1,449 EUR for Slovenia (Statistical Office of the Republic of Slovenia, 2024) and 803 EUR for Serbia (Statistical Office of the Republic of Serbia, 2024). The difference in the expected wages found through our research and the actual average wages in each country can be explained by not being familiar with the current average wage in each country, by an optimistic viewpoint of the students participating in our survey, and by a potential "delulu effect" (What does Delulu mean? | Later Social Media Glossary, 2024), which is eminent to individuals lacking real-world experience. The variety in expected wage increases after ten years of work can be clarified by the continuous, steady, and considerable annual increase in average salary in Serbia in recent years compared to Slovenia's more stable macroeconomic situation. Additionally, cultural background can also be identified as potential factor, as motivation towards achievement and success dimension of national culture is much higher in Serbia, so the students hope for faster career progress.

Through our research, we have found that the career and wage expectations among Gen Zers are connected. Whereas the themes per se are not expected to have a considerable influence in terms of wage expectations right after the completion of studies or starting to work, we can notice that in Slovenia, after ten years of work experience, such themes as Competition and Entrepreneurship have a statistically significant relation with wage expectations, while in Serbia this is the case with Organization Membership. However, in Serbia Entrepreneurship has a statistically significant relationship with starting wage expectations.

This study has several limitations that simultaneously expose several opportunities for further research. First, the discussed findings and implications were obtained from a single study; generalizing the results should be done cautiously. Second, the research mainly focused on expectations among Gen Z students. To have more valid results in separate analyses per country, we tend to widen the sample and repeat the research in future. Furthermore, the research can also include other countries.

Further research should examine the differences in career and wage expectations depending on different factors, such as gender, nationality, field of study, financial status, and others. Additionally, it would be interesting to see how the results change as respondents become more mature, for example, when they graduate and start working or after gaining several years of working experience.

References

Abbey, C.: The changing landscape of careers. Int. J. Law Manag. Humanit. **2**(4), 2922 (2021)
Alferjany, M.A.O.A., Alias, R.B.: Generational differences in values and attitudes within workplace. Psychol. Educ. J. **57**(9), 1496–1503 (2020)
Barhate, B., Dirani, K.M.: Career aspirations of Generation Z: A systematic literature review. Eur. J. Train. Dev. **46**(1/2), 139–157 (2022)
Blustein, D.L.: The Psychology of Working: A New Perspective for Career Development, Counseling, and Public Policy. Routledge (2006)
Brynjolfsson, E., McAfee, A.: The Second Machine Age: Work, Progress, and Prosperity in a Time of Brilliant Technologies. W.W. Norton & Company (2014)
Buchanan, J., Yu, S.: Career Guidance for Social Justice: Contesting Neoliberalism. Routledge (2018)
Career expectations questionnaire - career-expectations. pdf, 2024 https://www.birmingham.ac.uk/generic/decide/documents/career-expectations.pdf, Accessed 0 Apr 2024.
Deloitte (2020). Understanding Generation Z in the workplace: New employee engagement tactics for changing demographics. Deloitte https://www2.deloitte.com/us/en/pages/consumer-business/articles/understanding-generation-z-in-the-workplace. html, Accessed 14 Sep.2024.
Dolot, A.: Charact. Gener. Z. E-mentor. **2**(74), 44–50 (2018)
Fernandes, A., Huber, M., Vaccaro, G.: Gender differences in wage expectations. PLoS One. **16**(6), e0250892 (2021)
Fratričová, J., Kirchmayer, Z.: Barriers to work motivation of Generation Z. J. HRM. **21**(2), 28–39 (2018)
Friedman, T.L.: The World Is Flat: A Brief History of the Twenty-First Century. Farrar, Straus and Giroux, New York (2005)
Gabrielova, K., Buchko, A.A.: Here comes Generation Z: Millennials as managers. Bus. Horiz. **64**(4), 489–499 (2021)
Gaidhani, S., Arora, L., Sharma, B.K.: Understanding the attitude of generation Z towards workplace. Int. J. Manag. Technol. Eng. **9**(1), 2804–2812 (2019)
Geert Hofstede Country comparison graphs, 2024, https://geerthofstede.com/country-comparison-graphs/, last Accessed 20.Sep.2024
Gottfredson, L.S.: Applying gottfredson's theory of circumscription and compromise in career guidance and counseling. Career Dev. Q. **53**(3), 179–188 (2005)
Handshake (2020). Data Reveals Gen Z's Pragmatism About the Job Market & Student Debt. https://joinhandshake.com/blog/our-team/gen-z-pragmatism-about-the-coming-job-market-debt, last Accessed 14.Sept.2024.
Hofstede, G.: Culture and organizations. Int. stud. manag. organ. **10**(4), 15–41 (1980)
Janin, A.: Thanks to remote work, many in Gen Z may never work in an office. Will it matter. Wall Street J. https://www.wsj.com/articles/remote-lets-generation-z-work-without-any-office-will-matter-11645220523. (2022) Accessed 14 Sept 2024
Jenkins, R.: The Generation Z Guide: The Complete Manual to Understand, Recruit, and Lead the Next Generation. Lightning Source Inc (2019)
Kidd, J.M.: Exploring the components of career well-being and the emotions associated with significant career experiences. J. Career Dev. **35**(2), 166–186 (2008)
Kostić, S.C., Šarenac, J.G.: New normal strategic communication. In: Advances in Logistics, Operations, and Management Science Book Series, pp. 71–92 (2019)
Kuron, L.K., Lyons, S.T., Schweitzer, L., Ng, E.S.: Millennials' work values: Differences across the school to work transition. Pers. Rev. **44**(6), 991–1009 (2015)

Kovačević, V., Marič, M.: Upravljanje medgeneracijskega sodelovanja iz naslova predsodkov in izzivov. In: Šprajc, P., et al. (eds.) 43th International Conference on Organizational Science Development: Green and Digital Transition - Challenge or Opportunity: conference proceedings. University of Maribor, University Press, Maribor (2024)

Later Social Media Glossary, What does Delulu mean? 2024, https://later.com/social-media-glossary/delulu/, Accessed 11 Apr 2024.

Lent, R.W., Brown, S.D., Hackett, G.: Toward a unifying social cognitive theory of career and academic interest, choice, and performance. J. Vocat. Behav. **45**(1), 79–122 (1994)

Marič, M., Todorović, I., Marinčič, D.: Personal ethics perception and its relation to the expected properties of a good leader. In: Mihić, M., Jednak, S., Savić, G. (eds.) Sustainable Business Management and Digital Transformation: Challenges and Opportunities in the Post-COVID Era. Lecture Notes in Networks and Systems, vol. 562, pp. 595–605. Springer, Cham (2023)

Marič, M., Todorović, I., Žnidaršič, J.: Relations between work-life conflict, job satisfaction and life satisfaction among higher education lecturers. Manag.: J. Sustain. Bus. Manag. Solut. Emerg. Econ. **26**(1) (2021)

Maurer, R.: What HR should know about generation Z. Society for. Hum. Resour. Manag.. https://www.shrm.org/topics-tools/news/talent-acquisition/hr-know-generation-z. (2016) Accessed 15 Apr 2024

Mayrhofer, W., Steyrer, J., Meyer, M., Strunk, G., Schiffinger, M., Iellatchitch, A.: Graduates' career aspirations and individual characteristics. Hum. Resour. Manag. J. **15**(1), 38–56 (2005)

Milinković, I., Kovačević, I., Mihailović, D. What Do Freshmen Want? Career Path Preferences Among Students. Manag.: J. Sustain. Bus. Manag. Solut. Emerg. Econ., 22(1), 37–45 (2017).

Nedeljković, M., Vukonjanski, J., Nikolić, M., Hadžić, O., Šljukić, M.: A comparative analysis of Serbian national culture and national cultures of some European countries by GLOBE project approach. J. Geogr. Inst. "Jovan Cvijic" SASA. **68**(3), 363–382 (2018)

Nguyen Ngoc, T., Viet Dung, M., Rowley, C., Pejić Bach, M.: Generation Z job seekers' expectations and their job pursuit intention: Evidence from transition and emerging economy. Int. J. Eng. Bus. Manag. **14**, 18479790221112548 (2022)

Okanović, M., Milosavljević, M., Cicvarić Kostić, S., Dlačić, J.: Embryonic tech ventures' orientation and performance. Manag.: J. Sustain. Bus. Manag. Solut. Emerg. Econ. (2023)

Pauli, U., Buchelt, B., Pocztowski, A.: Self-employment: A promising agenda for IHRM research. Eur. J. Int. Manag. **13**(1), 111–125 (2019)

Rakić, J.Ž., Krstić, G., Oruč, N., Bartlett, W.: Income inequality in transition economies: A comparative analysis of Croatia. Serb. Slov., Econ. Ann. **64**(223), 39–60 (2019)

Saha, L.J., Sikora, J.: The career aspirations and expectations of school students: from individual to global effects. Educ. Soc. **26**(2), 5–22 (2008)

Savickas, M.L.: Career Construction: A Developmental Theory of Vocational Behavior. In: Career Choice and Development, 4th edn. Brown & Associates, John Wiley & Sons (2002)

Seemiller, C., Grace, M.: Generation Z: A century in the making. Routledge (2018)

Silva, J., Carvalho, A.: The work values of Portuguese Generation Z in the higher education-to-work transition phase. Soc. Sci. **10**(8), 297 (2021)

Singh, A.P., Dangmei, J.: Understanding the generation Z: the future workforce. South-Asian J. Multidiscip. Stud. **3**(3), 1–5 (2016)

Statistical Office of the Republic of Serbia (2024). https://www.stat.gov.rs/sr-latn/vesti/statisticalrelease/?p=15103&a=24&s=2403?s=2403., Accessed 01 May 2024.

Statistical Office of the Republic of Slovenia (2024). Plače in stroški dela. https://www.stat.si/statweb/Field/Index/15., Accessed 01 May 2024.

Tkalac Verčič, A., Verčič, D.: Attitudes toward sustainable development and employer brands: Comparing generations X, Y and Z in two countries. Corp. Commun.: Int. J. **30**, 355–371 (2024)

Todorović, I., Okanović, M., Cicvarić Kostić, S., Marič, M., Pihir, I.: Entrepreneurial Mindset of Students in Central and Eastern Europe: Factors that Determine Intentions and Actions. In: Journal of East European Management Studies special issue 2024 on Entrepreneurship Education, paper accepted for publishing in the Special Issue, pp. 112–135 (2024)
Twenge, J.M.: Generation Me: Why Today's Young Americans Are More Confident, Assertive, Entitled—and More Miserable than Ever Before. Simon and Schuster (2010)
Twenge, J.M.: IGen: Why Today's Super-Connected Kids are Growing up Less Rebellious, More Tolerant, Less Happy--and Completely Unprepared for Adulthood--and What That Means for the Rest of us. Simon and Schuster (2017)
Vesić, G., Arsenijević, G., Marič, M.: Digital Political Marketing. Pearson, Harlow (2024)
Vroom, V.H.: Work and motivation. Wiley, New York (1964)

AI-Powered Onboarding Process: A Gamechanger for Onboarding Success and Outcomes

Ana Belajdžić[1(✉)] and Sara Delac Marković[2]

[1] Department of Psychology, Faculty of Humanities and Social Sciences, Josip Juraj Strossmayer University of Osijek, Osijek, Croatia
abelajdzic@ffos.hr
[2] Infobip, Zagreb, Croatia

Abstract. Effective recruitment and onboarding are crucial for optimizing new hire performance and organizational success in the contemporary business landscape. This research compares the effect of traditional versus **AI-powered onboarding** processes on new hire performance through the lens of procedural justice and onboarding success. The research comprises two distinct studies conducted in a private-sector organization within the EU. Study 1 was carried out with N = 154 new hires who underwent **traditional onboarding**, while Study 2 included N = 112 new hires who went through an AI-powered onboarding. Mediation analysis revealed that in traditional onboarding, perceived procedural justice directly enhances **new hire performance**, with supervisors' justice affecting performance evaluations independently of onboarding success. Conversely, in AI-powered onboarding, **procedural justice** affects performance both directly and indirectly through perceived onboarding success. The latter indicates that AI-driven processes integrate procedural justice and onboarding success, yielding improved performance outcomes. Specifically, new hires who viewed the AI onboarding process as procedurally fair performed better, partly due to their perception of onboarding success. These findings underscore notable differences between hire traditional and AI-powered onboarding methods. Traditional onboarding relies on direct supervisory justice to boost performance, whereas AI-powered onboarding integrates procedural justice with onboarding success, leading to more effective outcomes. The study contributes valuable insights into how onboarding practices can be enhanced through AI, emphasizing the importance of procedural justice and its role in optimizing new hire performance. Organizations may acquire superior performance outcomes by adopting AI technologies in their onboarding processes.

Keywords: AI-powered onboarding success · procedural justice · new hire performance

V. Štavljanin et al. (Eds.): SymOrg 2024, LNNS 1681, pp. 123–135, 2025.
https://doi.org/10.1007/978-3-032-08090-5_10

1 Introduction

Onboarding is such an essential process as it helps shape the new hire's experience from day one and impacts their long-term success within organizations [26]. Traditionally, onboarding depended extensively on manual procedures and face-to-face systems, considerably reducing control and efficiency [22]. As technology has advanced, AI technology opens new horizons that one can utilize to augment the onboarding experience. More recently, AI-powered onboarding systems advanced to promise experiences that were increasingly personalized and efficient – effectively revolutionizing the way we have come to do things [18, 30].

Given the importance of preserving employee perceptions of **organizational justice**, it may be worthwhile to investigate whether AI integration alters the fairness employees perceive in organizational processes and decisions [24, 31]. This is important because, in traditional **onboarding processes**, subjective evaluations and inconsistent application of policies could lead to perceived unfair treatment, thereby impacting perceptions of fairness and satisfaction [23]. The data-driven nature of AI-powered onboarding promises to create a more equal procedure and distribution of rewards, which may improve procedural and distributive justice perceptions [9].

Also, new employees are generally only as successful as their onboarding, a key driver of new hire performance and future tenure [10]. It is critical to appreciate how effective onboarding results in higher job satisfaction and faster, better integration with performance outcomes [14]. This will help improve these aspects more effectively than traditional methods [31] and through real-time feedback from personalized learning that AI-driven systems could provide. New hires who perceive higher procedural justice of the AI-powered onboarding process tend to perform better, partially because they perceive the onboarding process as successful [3]. However, to our knowledge, the comparative effect of traditional and AI-driven onboarding systems on onboarding success and its outcomes has been underexplored. Therefore, this research aims to compare traditional and AI-powered onboarding processes, in the light of the relationship between procedural justice, onboarding success and new hire performance.

2 Literature Review

The ultimate goal of an organization which they always seek for is the high performance of new hires [8]. In today's dynamic business environment, the moments of recruiting and integrating new employees are decisive for organizational performance. These new hires will have a major impact on company performance. For one, new hires directly impact organizational productivity and efficiency by virtue of their performance. As they adjust to what their position requires of them, how prompt they are in understanding tasks, coordinating with team members, and adding value to the projects can either be beneficial or damage organizational productivity. New hires who excel can drive team productivity, developing and creating new innovative solutions, whereas those who fail to perform can hinder this progress and drain resources [5].

Identifying these high-quality candidates is just the start of a much longer integration process. Finally, the gap that separates new hires from swiveling from heart of an organization is filled by powerful onboarding processes, establishing incoming employee performance and paving the way for their success in the company. This process includes a set of planned activities and experiences that help to integrate new hires into the organization, facilitate their **socialization** and practice as well as offer the information, instructions or guidance needed to perform successfully. Orientation to range of integration (culture, job, expectations and relationship-building support training) involves gathering new information on a series of aspects of the organization that an individual will be joining [1]. According to research, properly designed onboarding programs lead to increased job satisfaction, organizational commitment and productivity in newly hired employees [4]. Moreover, onboarding aids in the socialization of new employees into an organizational culture which helps them feel more integrated and identified with the company values [5]. New research further emphasizes the need for customized onboarding strategies that include mentorship programs and mechanisms to provide employees with feedback, with more of an eye toward talent management within these early months [5]. With these strategies, organizations can streamline the implementation of new hires to full-scale productivity within the organization, thereby enhancing their ability to contribute towards meeting organizational goals and success. In recent years, we have also come to realize that incorporating onboarding tools and applications is proving to be revolutionary, as it provides a competitive advantage to boost the efficiency of traditional methods involved in onboarding. These technologies automate administrative tasks and provide tailored training content and instant feedback, all to fast-track the learning curve and smooth out changes for new staff [12]. Using AI-driven insights and predictive analytics, organizations can discover gaps sooner and personalize the onboarding experience to individual needs, making orientation more successful [9].

However, in this digital revolution, the beacon that helps provide an anchor for us is a return to procedural justice. Procedural justice — the extent to which procedures adopted by organizations are viewed as fair, transparent, and consistent (in determining outcomes) in organizational contexts—has received substantial attention in the Organizational Psychology literature due to its ramifications on employee attitudes, behaviors, and outcomes [6]. Procedural justice is a strong predictor of employee resilience during challenging organizational changes, such as integration [15]. It also enhances employee engagement [33] and improves team effectiveness evaluations [29]. Furthermore, procedural justice arises as a critical driver of new hire performance and its implications for the perception of fairness or organizational commitment in the onboarding process. For example, research has found that procedural justice during onboarding strengthens the alignment between new hires and their organization [19], leading to greater trust in organizations and less uncertainty. The latter will help employers to have clear and consistent steps, which in turn gives confidence to their new hires and, therefore, higher engagement and job satisfaction [6]. In addition, procedural justice at onboarding is important for a positive organizational climate, cooperation, and teamwork [28]. Empirical studies have underscored how targeted onboarding practices improve procedural justice and new hire performance. Tailored communication strategies with clear expectations and voice/feedback options mobilize new employee, which, in turn, support their cognition about organizational justice and procedural fairness alignment within the organization [21]. At the same time, harnessing technology-powered platforms and AI-based tools in onboarding provides a chance to make processes more uniform by maintaining consistency and **transparency**, adding another layer of support for procedural justice [12, 17].

How a new employee feels about onboarding is a direct reflection of whether they will be successful in the future within an organization. Research has suggested that favorable perceptions of onboarding success correlate with increased levels of job success, organizational commitment, and satisfaction among valued newcomers [1, 32]. The outcome of conservative onboarding is that new hires feel valued or as if they are truly aligned with the organization's strategies and goals. Emerging research explores critical drivers of onboarding success as determined by new employees. The extent of the clear communication provided and the training and resources given to employees also created their attitudes toward onboarding [22]. In addition, when onboarding activities are consistent with the organization's culture and values during new employee assimilation, the new hires become more likely to believe in successful onboarding [32]. Moreover, it is essential to acknowledge that implementing technology-enabled solutions has improved new hires' perception of onboarding success. With the help of AI-powered platforms, virtual reality simulations and mobile applications can further be strengthened to personalize and significantly amplify the onboarding experience, making new hires happier with their perceived level of readiness [32]. By using technology to smooth out the process and offer timely support and culture-building connections with colleagues, organizations can create positive impressions of onboarding success that promote high new hire performance.

This research can supplement the discussion of changes in onboarding today by comparing interaction design processes that apply AI to traditional approaches. The latter would advance our theoretical understanding of procedural justice in contemporary organizational contexts, provide empirical insight into the effectiveness of onboarding practices on new hire success, and assess performance gains resulting from AI-driven processes. Ultimately, this comparison can serve as a blueprint for designing fairer, more effective — and ultimately performance-improving onboarding programs.

3 Method

3.1 Participants and Procedure

This research, comprising two distinct studies, was conducted in the context of consulting services provided to a private-sector organization within an EU member state.

Study 1 involved a sample of N = 154 new hires who underwent a traditional onboarding process, wherein their immediate supervisors assisted them in acclimating to their new roles and addressing any inquiries over a three-month (90-day) period. The majority of participants were male (51%), with educational backgrounds ranging from high school diplomas (20%) to Bachelor's (17%) and Master's degrees or higher (63%). Upon completion of the onboarding process, the new hires provided evaluations of their supervisors' procedural justice and the overall success of their onboarding experience using a paper-and-pencil method. Concurrently, their supervisors assessed the new hires' job performance.

Study 2 included a sample of N = 112 new hires who participated in an AI-powered onboarding process. The demographic profile was similar, with 52% of participants being male, and educational qualifications of either high school diplomas (37%) or Bachelor's degrees and higher (63%). The organization implemented an AI-powered application to facilitate the onboarding process, aiming to improve its efficiency and the overall experience for the recruits. This AI-driven process also spanned three months (90 days). Afterward, the new hires assessed the procedural justice and the success of their onboarding experience, using a paper-and-pencil method during final workshops, while their supervisors provided performance evaluations.

The data for both studies were collected over a 19-month period and analyzed using Hayes PROCESS Macro (2018, Model 4) within the IBM SPSS Statistics 25 software package.

3.2 Instruments

The same measurement instruments were utilized in both studies, with the *Procedural justice scale* being slightly adapted to assess perceptions of a target-specific justice source (i.e., supervisor in Study 1 or the AI-powered onboarding tool in Study 2).

Procedural justice was measured using a five-item scale developed by Jakopec & Sušanj (2014). In Study 1, sample items included: "My supervisor provides me with clear feedback on the quality of my work" and "My supervisor explains decisions clearly and offers additional information when requested." In Study 2, the scale was modified to assess perceptions of procedural justice in the AI-powered onboarding process. Equivalent items for this study included: "The AI-powered onboarding tool provides me with clear feedback on the quality of my work" and "The AI-powered onboarding tool explains decisions clearly and offers additional information when requested." Participants rated each item on a five-point Likert scale, ranging from 1 (strongly disagree) to 5 (strongly agree), where higher scores indicated greater perceived procedural justice from supervisors (Study 1) or the AI-powered system (Study 2).

Onboarding success was assessed using a five-item scale by Sharma & Stol (2019). Example items included: "The initial orientation program helped me feel less stressed about joining a new workplace" and "I clearly understand the expectations and responsibilities of my job." These items were similarly rated on a five-point Likert scale, with higher scores indicating greater perceived success in the onboarding process.

New hire performance was evaluated through supervisor ratings, with performance criteria established at the beginning of the onboarding period. At the conclusion of the onboarding process, supervisors rated the new hires' performance on a five-point scale: "unsatisfactory," "requires attention," "satisfactory," "above satisfactory," and "excellent."

4 Results and Discussion

As presented within Table 1, new hires who underwent a traditional onboarding process (Study 1) perceive their immediate supervisor as fair in terms of procedural justice. They feel that their supervisor provides them with explicit feedback on the quality of their work, explains decisions clearly, and offers additional information when requested. Similarly, new hires who participated in an AI-powered onboarding process (Study 2) feel the same regarding the procedural fairness of the AI-powered onboarding process. Although they perceive their onboarding process as successful and feel that the initial orientation program helped them feel less stressed about joining a new workplace, new hires that underwent an AI-powered onboarding process are more critical while observing the overall onboarding success compared to the new hires that participated in a traditional onboarding process, possibly indicating higher expectations from AI systems or more precise observations facilitated by AI-driven data and feedback mechanisms. Users often expect AI systems to minimize human error, and when these systems fall short, the disparity between expectations and reality can result in more critical evaluations [13]. Supervisors' ratings of the new hires' performance are slightly higher within the traditional onboarding process as compared to the AI-powered onboarding process. This may indicate that the more human-centered approach in traditional onboarding helps build stronger initial supervisor-employee relationships, which could positively affect performance ratings. Interpersonal interactions between supervisors and new hires during traditional onboarding play a crucial role in fostering trust, rapport, and social support [2]. These strong initial relationships can lead to better communication, role clarity, and performance outcomes.

Table 1. Descriptive statistics of all variables measured in Study 1 (N = 154) and Study 2 (N = 112).

Variables	Descriptive statistics Study 1			Descriptive statistics Study 2		
	M	*SD*	*α*	*M*	*SD*	*α*
Procedural justice	4.05	0.73	.96	4.08	1.13	.94
Onboarding success	3.50	1.05	.88	3.42	1.03	.83
New hire performance	4.55	0.50	.86	4.25	1.02	.88

As expected, perceived procedural justice strengthens new hires' evaluations of onboarding success and their performance rated by their immediate supervisors. Additionally, onboarding success leads to higher ratings of new hires' performance. These relations, presented within Table 2, are even more pronounced in the case of AI-powered onboarding process (Study 2) compared to the traditional onboarding process (Study 1). Despite descriptive statistics being more on the side of traditional methods, the relations between perceived procedural justice, onboarding success and new hire performance are significantly stronger in the AI-powered onboarding process, suggesting that AI may reinforce perceptions of fairness and transparency more effectively than traditional methods, ultimately leading to more pronounced effects on onboarding success and job performance. Research suggests that AI systems enhance procedural justice by providing greater consistency, transparency, and objectivity in decision-making processes, which can positively affect employees' perceptions of justice [27].

Table 2. Intercorrelations of all variables measured in Study 1 (N = 154) and Study 2 (N = 112).

Variables	Intercorrelations Study 1			Intercorrelations Study 2		
	1	2	3	1	2	3
1. Procedural justice	–	.48**	.28**	–	.69**	.66**
2. Onboarding success		–	.27**		–	.59**
3. New hire performance			–			–

Note: $^{**}p < .01$

The mediation analyses results (depicted in Fig. 1 and Fig. 2) support the above-mentioned. As depicted in Fig. 1, new hires who underwent a traditional onboarding process perceive their immediate supervisor as fair in terms of procedural justice, which makes them perceive the overall onboarding process successful [LLCI 0.4978, ULCI 0.9018]. Perceived onboarding success leads to higher supervisors' rating of new hires' performance [LLCI 0.0008, ULCI 0.1677]. However, when it comes to the traditional onboarding process, procedural justice positively affects new hires' performance only directly. In other words, new hires who underwent a traditional onboarding process and perceive their supervisor as fair with respect to procedural justice, receive higher performance evaluations from their supervisors [LLCI 0.0149, ULCI 0.2556]. The indirect effect of perceived procedural justice on new hires' performance (via onboarding success) is insignificant [LLCI - 0.0015, ULCI 0.1238], meaning that the new hires' perceptions of onboarding success are not the reason why procedural justice positively affects the performance of new hires' that underwent the traditional onboarding process. To conclude, the mediation analysis results depicted in Fig. 1 suggest that in the traditional onboarding process, perceived procedural justice directly affects new hires' performance. New hires who perceive their supervisor as fair tend to receive higher performance evaluations, independent of their perceptions of onboarding success. While perceived onboarding success positively affects performance, it does not mediate the relationship between procedural justice and performance. In other words, procedural fairness in supervision leads directly to better performance outcomes, with onboarding success playing a separate but less significant role in this process. The tested model explained 47% of the outcome variance ($p < .001$).

However, when it comes to the AI-powered onboarding process (Study 2), results of the mediation analysis somewhat differ. Similarly, as new hires who underwent a traditional onboarding process, the new hires that underwent the AI-powered onboarding process perceive the AI-powered onboarding process fair in terms of procedural justice, which makes them perceive the overall onboarding process successful [LLCI 0.5071, ULCI 0.7588]. Perceived onboarding success further leads to higher supervisors' rating of new hires' performance [LLCI 0.0723, ULCI 0.4508]. Unlike it was the case with the traditional onboarding process, when it comes to the AI-powered one, procedural justice positively affects new hires' performance both directly, and indirectly, via onboarding success. In other words, new hires who underwent the AI-powered onboarding process and perceive the onboarding process procedurally fair, receive higher performance evaluations from their supervisors [LLCI 0.2613, ULCI 0.6090]. However, in the case of

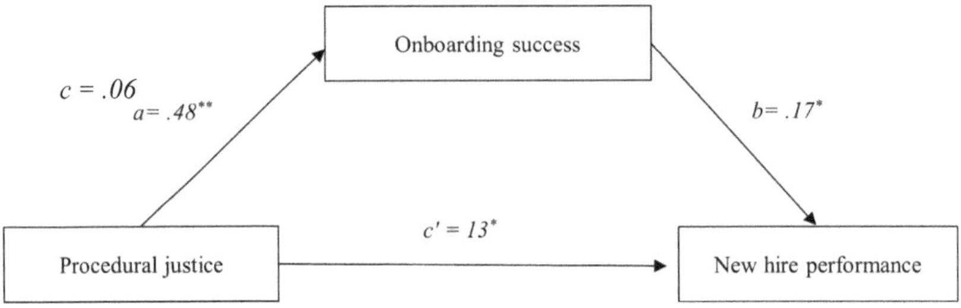

*Note: a, b – standardized regression coefficients; c' – direct effect; c – indirect effect; *p < .05*

Fig. 1. Graphic representation of the mediating effect of onboarding success in the relationship between procedural justice and new hire performance (Study 1; N = 154).

AI-powered onboarding process, the indirect effect of perceived procedural justice on new hires' performance (via onboarding success) is significant [LLCI 0.0458, ULCI 0.3434]. Namely, procedural justice both, directly and indirectly – via onboarding success – strengthens new hires' performance. In summary, new hires that perceive higher levels of procedural justice of the AI-powered onboarding process tend to perform better, partially because they perceive the onboarding process successful. The tested model depicted in Fig. 2 explained 68% of the outcome variance ($p < .001$).

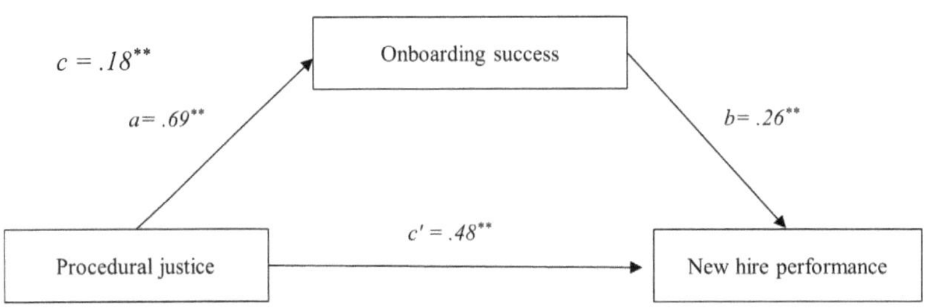

*Note: a, b – standardized regression coefficients; c' – direct effect; c – indirect effect; **p < .01*

Fig. 2. Graphic representation of the mediating effect of onboarding success in the relationship between procedural justice and new hire performance (Study 2; N = 112).

The results above highlight significant differences between the traditional and AI-powered onboarding processes regarding the relationship between procedural justice, onboarding success, and new hire performance.

In the traditional onboarding process, perceived procedural justice directly affects new hires' performance. New hires who perceive their supervisor as fair receive higher performance evaluations from their supervisors. The direct relationship between perceived procedural justice and performance aligns with previous research demonstrating that fair treatment from supervisors fosters positive employee outcomes, such as job performance [6, 29, 33]. While onboarding success positively affects performance, it does not mediate the relationship between procedural justice and new hire performance. In other words, procedural justice directly enhances performance, with onboarding success playing a secondary role. The lack of a mediating effect of onboarding success suggests that, in traditional settings, fairness perceptions toward the supervisor may overshadow other onboarding factors. This is consistent with research indicating that interpersonal interactions during onboarding, such as supervisor-employee relationships, are critical to performance [1]. In these contexts, procedural justice directly affects performance without relying on other onboarding elements to mediate this relationship. The tested model explained 47% of the variance in performance outcomes, indicating that other factors also contribute to performance in the traditional onboarding process, which is in line with findings that suggest multiple factors beyond onboarding success, such as social integration and organizational culture, contribute to new hires' performance [20].

In contrast, the AI-powered onboarding process exerts both, direct and indirect effects of procedural justice on performance. Perceived procedural justice positively affects new hires' performance directly and also indirectly via onboarding success, which suggests that in AI-powered onboarding, procedural justice not only improves performance on its own but also strengthens new hires' perceptions of onboarding success, which further enhances performance. AI systems often provide greater consistency and transparency in decision-making and feedback, which can enhance perceptions of fairness [7]. The structured nature of AI-driven processes may lead new hires to perceive the onboarding as more successful, amplifying its impact on performance. This aligns with studies showing that AI-driven processes enhance procedural justice by providing clear, data-driven explanations and removing potential human biases [27]. The indirect effect of procedural justice through onboarding success supports evidence that technology-enabled onboarding improves job readiness and performance by ensuring that new hires are better supported and integrated into the organization [25].

The higher explanatory power of the AI-powered onboarding model (68%) indicates that AI-driven processes may be more effective in addressing both the procedural fairness and success of the onboarding experience, which further supports research suggesting that digital tools can improve onboarding outcomes through enhanced personalization and feedback [2].

The evidence suggests that traditional onboarding heavily relies on the direct effect of procedural justice from supervisors to enhance new hires' performance. In contrast, AI-powered onboarding processes integrate procedural justice with onboarding success, resulting in more effective and cohesive onboarding outcomes. The AI-driven approach aligns with literature highlighting the advantages of AI in improving transparency, fairness, and feedback consistency, thereby promoting better performance [11]. As AI continues to be integrated into HR processes, organizations may benefit from its ability to enhance both procedural justice and the overall onboarding experience.

The scientific value of these research findings lies in their contribution to our understanding of how different onboarding methods affect new hire performance through the lens of procedural justice and onboarding success. Traditional onboarding emphasizes direct supervisory justice, while AI-powered onboarding enhances both procedural justice and onboarding success through its structured and transparent approach. This not only reinforces the importance of procedural justice but also demonstrates the potential for AI to improve onboarding processes and outcomes. The results suggest that organizations could achieve more effective onboarding and better performance outcomes by integrating AI technologies into their onboarding practices.

However, future research should address the existing research limitations. The cross-sectional nature of the data and reliance on correlational analyses preclude causal inferences, and alternative statistical techniques, such as longitudinal or experimental designs, could offer further clarity on the causal mechanisms underlying these relationships. Longitudinal studies tracking changes in performance and perceptions over time could clarify causal mechanisms, while randomized controlled trials or matched samples could help isolate the effects of onboarding methods. Additionally, reliance on self-report measures for onboarding success may introduce response bias or social desirability effects, potentially influencing the accuracy of the results. Future research could complement self-reports with objective measures of onboarding success, such as employee retention rates, performance evaluations, or time-to-productivity metrics, to provide more reliable indicators of onboarding effectiveness. The relatively short duration of the onboarding process (90 days) may constrain the ability to fully evaluate the long-term impact of AI-powered onboarding on new hires' perceptions and performance. While initial outcomes, such as early productivity and procedural justice perceptions, can be observed within this timeframe, it remains unclear how these effects persist or evolve as employees integrate further into the organization. Longer-term follow-up assessments, potentially spanning six months to a year or more, could offer valuable insights into whether the benefits of AI-powered onboarding are maintained, diminish over time, or lead to additional outcomes, such as career progression or enhanced organizational commitment. Such longitudinal data would provide a more holistic understanding of the AI onboarding process's effectiveness and its role in shaping long-term employee success and organizational impact. Further, AI-powered onboarding offers significant advantages but also presents challenges. A key concern is job displacement, particularly among HR staff, as automation may reduce opportunities for human-centered tasks. Additionally, implementing AI requires training employees to adapt to new technologies, which could create a steep learning curve and widen inequalities for those with limited technological skills. There are also risks of bias from flawed data or algorithms, which may disadvantage certain groups. Over-reliance on AI may reduce human connection and cultural integration during onboarding. Privacy and ethical concerns around data collection and decision-making could pose compliance challenges. To address these, organizations should balance automation with human oversight, invest in training, audit AI for bias, and ensure transparency and data security. Finally, the study's focus on a single private-sector organization limits its generalizability. Future research should explore AI-powered onboarding across various industries, including public-sector organizations, non-profits, and multinational corporations, to assess differences in implementation and outcomes. Examining organizations of different sizes and across cultural or geographic contexts could reveal how resource availability and employee expectations shape AI adoption. Broadening the scope would provide a more nuanced understanding of AI-powered onboarding's potential and its adaptability to diverse organizational settings.

Nevertheless, study findings offer valuable practical insights for organizations. They highlight the advantages of adopting AI-driven onboarding tools that provide structured, data-driven feedback and maintain consistency. Organizations should also invest in training programs to equip supervisors with the skills needed to deliver fair and constructive feedback. A balanced approach that integrates AI tools with human oversight can leverage the strengths of both methods. Additionally, organizations should use data from AI systems to continuously refine and enhance the onboarding process. It is crucial that both AI-driven and traditional onboarding methods emphasize fairness and transparency. To ensure ongoing effectiveness, organizations should regularly assess the onboarding programs and adjust strategies based on feedback and performance outcomes.

Declaration of AI and AI-Assisted Technologies in the Writing Process. *During the preparation of this work the authors used ChatGPT and Grammarly for streamlining, proofreading and English language editing purposes. After using these tools, the authors reviewed and edited the content as needed and take full responsibility for the content of the publication.*

References

1. Bauer, T.N., Bodner, T., Erdogan, B., Truxillo, D.M., Tucker, J.S.: Newcomer adjustment during organizational socialization: a meta-analytic review of antecedents, outcomes, and methods. J. Appl. Psychol. **92**(3), 707 (2007)
2. Bauer, T.N., Erdogan, B.: Organizational socialization: The effective onboarding of new employees. In: Zedeck, S. (ed.) APA Handbook of Industrial and Organizational Psychology, Maintaining, Expanding, and Contracting the Organization, vol. 3, pp. 51–64. American Psychological Association (2011)
3. Belajdžić, A.: The power of AI onboarding: Fostering fairness, onboarding success and performance in new hires. In: SymOrg 2024 Xix International Symposium, pp. 731–735 (2024)
4. Bell, T.: Onboarding: Improving employer and employee relations. Certif. Public Manag. Appl. Res. **2**(1), 1 (2021)
5. Chowdhury, S., Budhwar, P., Dey, P.K., Joel-Edgar, S., Abadie, A.: AI-employee collaboration and business performance: Integrating knowledge-based view, socio-technical systems and organizational socialization framework. J. Bus. Res. **144**, 31–49 (2022)
6. Colquitt, J.A., et al.: Justice at the millennium, a decade later: A meta-analytic test of social exchange and affect-based perspectives. J. Appl. Psychol. **98**(2), 199 (2013)
7. Duan, Y., Edwards, J.S., Dwivedi, Y.K.: Artificial intelligence for decision making in the era of Big Data–evolution, challenges and research agenda. Int. J. Inf. Manag. **48**, 63–71 (2019)
8. Dutta, D., Vedak, C., Joseph, V.: New hire performance: Identifying critical traits for hospitality industry. Tour. Rev. **79**(1), 69–84 (2024)
9. Gélinas, D., Sadreddin, A., Vahidov, R.: Artificial intelligence in human resources management: A review and research agenda. Pac. Asia J. Assoc. Inf. Syst. **14**(6), 1 (2022)
10. Ghani, B., et al.: Challenges and strategies for employee retention in the hospitality industry: A review. Sustainability. **14**(5), 2885 (2022)
11. Glikson, E., Woolley, A.W.: Human trust in artificial intelligence: Review of empirical research. Acad. Manag. Ann. **14**(2), 627–660 (2020)

12. Gupta, C.P., Kumar, V.R.: Artificial intelligence application in human resource management: the way forward. In: 2024 ASU International Conference in Emerging Technologies for Sustainability and Intelligent Systems (ICETSIS), pp. 1726–1730. Institute Electrical and Electronics Engineers, Manama, Bahrain (2024)
13. Guzman, A.L., Lewis, S.C.: Artificial intelligence and communication: A human–machine communication research agenda. New Media Soc. **22**(1), 70–86 (2020)
14. Houle, S.A., Morin, A.J., Vandenberghe, C., Tóth-Király, I., Fernet, C.: From employee socialization to co-evolution: A lifespan multidisciplinary conceptualization. Appl. Psychol. **1**(41) (2024)
15. Jakopec, A.: Employee resilience during organizational integration: (Non)financial rewards and the mediating role of organizational justice. Suvrem. Psihol. **24**(2), 177–188 (2021)
16. Jakopec, A., Sušanj, Z.: Verifying the dimensionality of justice construct in organizational context. Psihol. teme. **23**(2), 305–325 (2014)
17. Jiang, L., Qin, X., Yam, K.C., Dong, X., Liao, W., Chen, C.: Who should be first? How and when AI-human order influences procedural justice in a multistage decision-making process. PLoS One. **18**(7), e0284840 (2023)
18. Johnson, R.D., Stone, D.L., Lukaszewski, K.M.: The benefits of eHRM and AI for talent acquisition. J. Tour. Futures. **7**(1), 40–52 (2020)
19. Jordan, S.L., Palmer, J.C., Daniels, S.R., Hochwarter, W.A., Perrewé, P.L., Ferris, G.R.: Subjectivity in fairness perceptions: How heuristics and self-efficacy shape the fairness expectations and perceptions of organizational newcomers. Appl. Psychol. **71**(1), 103–128 (2022)
20. Klein, H.J., Polin, B.: 14 Are organizations on board with best practices onboarding? Oxf. handb. organ. social. **267** (2012)
21. Korkmaz, A.V., Van Engen, M.L., Knappert, L., Schalk, R.: About and beyond leading uniqueness and belongingness: A systematic review of inclusive leadership research. Hum. Resour. Manag. Rev. **32**(4), 100894 (2022)
22. Liu, S., Watts, D., Feng, J., Wu, Y., Yin, J.: Unpacking the effects of socialization programs on newcomer retention: A meta-analytic review of field experiments. Psychol. Bull. **150**(1), 1 (2024)
23. Mer, A.: Artificial intelligence in human resource management: Recent trends and research agenda. Digit. Transform. Strateg. Resil. Cyber Secur. Risk Manag. **111**, 31–56 (2023)
24. Narayanan, D., Nagpal, M., McGuire, J., Schweitzer, S., De Cremer, D.: Fairness perceptions of artificial intelligence: A review and path forward. Int. J. Hum.–Comput. Interact. **40**(1), 4–23 (2024)
25. Sani, K.F., Adisa, T.A., Adekoya, O.D., Oruh, E.S.: Digital onboarding and employee outcomes: empirical evidence from the UK. Manag. Decis. **61**(3), 637–654 (2023)
26. Sharma, G.G., Stol, K.J.: Exploring onboarding success, organizational fit, and turnover intention of software professionals. J. Syst. Softw. **159**, 110442 (2020)
27. Shrestha, Y.R., Ben-Menahem, S.M., Von Krogh, G.: Organizational decision-making structures in the age of artificial intelligence. Calif. Manag. Rev. **61**(4), 66–83 (2019)
28. Smith, C.E., Matthews, R.A., Mills, M.J., Hong, Y.H., Sim, S.: Organizational benefits of onboarding contingent workers: An anchoring model approach. J. Bus. Psychol. **37**(3), 525–541 (2022)
29. Sušanj, Z., Jakopec, A.: Towards an understanding of the relationship between supervisors felt trust and team-effectiveness evaluation. Primenj. psihol. **9**(4), 413–427 (2016)
30. Votto, A.M., Valecha, R., Najafirad, P., Rao, H.R.: Artificial intelligence in tactical human resource management: A systematic literature review. Int. J. Inf. Manag. Data Insights. **1**(2), 100047 (2021)

31. Zhang, X., Antwi-Afari, M.F., Zhang, Y., Xing, X.: The impact of artificial intelligence on organizational justice and project performance: A systematic literature and science mapping review. Buildings. **14**(1), 259 (2024)
32. Zhao, T., Liu, J., Zawacki, A.M., Michel, J.S., Li, H.: The effects of newcomer proactive behaviors on socialization outcomes: A meta-analysis. J. Occup. Organ. Psychol. **96**(1), 1–32 (2023)
33. Zlatić, L., Sušanj, Z., Jakopec, A.: The role of perceived justice, external prestige and organizational identification in explaining employee engagement. Društvena istraživanja časopis za opća društvena pitanja **30**(3), 571–592 (2021)

The Role of Strategic HRM in Change Leadership: Driving Organizational Transformation

Milena Marjanović(✉) and Marko Mihić

Faculty of Organizational Sciences, University of Belgrade, 11000 Belgrade, Republic of Serbia
milena.marjanovic.office@gmail.com

Abstract. This study presents an analysis of the **organizational change management model** and the impact of **strategic human resource management** on **change leadership** and the organizational change management model. The methodology used involves case studies of acquisitions in supply chains within the markets of Central and Southeast Europe. The goal of this study is to develop a theoretical conceptual model of organizational change management through **human resource management** and change leadership during company transformations, with the purpose of increasing the efficiency and effectiveness of organizational change management. The findings suggest that there is a high level of influence between organizational change management and modern strategic human resource management during **transformations**, which further contributes to the development of both the theoretical and practical concepts of organizational change management. This study provides an overview of the development of the organizational change management model, offering a deeper understanding, guidance for decision-makers, **leaders**, and managers, and encouraging further research by scholars. The limitations of this study include its descriptive approach, the limited availability of literature, and the small sample size used in the case studies.

Keywords: Change Management Model · Strategic Human Resource Management · Change Leadership

1 Introduction

Today's business environment and company management face numerous challenges, starting with technological innovations and, based on these, the development of new professions, management platforms, and models (Petrović et al., 2008). The goal of this study is to develop a theoretical conceptual model of organizational change management through human resource management and change leadership during company transformations, with the purpose of increasing the efficiency and effectiveness of organizational change management. The scientific goal of this research is to provide a systematic review of the available scientific literature in the areas of organizational change management, strategic human resource management, and leadership. This research highlights the link

V. Štavljanin et al. (Eds.): SymOrg 2024, LNNS 1681, pp. 136–165, 2025.
https://doi.org/10.1007/978-3-032-08090-5_11

between organizational change management and strategic human resource management during transformations, which further contributes to the development of both the theoretical and practical concepts of organizational change management. The findings suggest that there is a high level of influence of human resource management strategies on organizational change models. This study provides an overview of the development of organizational change management models, offering guidance for decision-makers, leaders, and managers.

Strategies for managing organizational changes and human resource management strategies from the perspective of corporate social responsibility (CSR) were discussed, through the contribution of the company's conceptual strategic directions to the social community, such as the utilization of local and global workforces, based on economic, legal, and ethical responsibilities (Mihic et al., 2019). Additionally, the paper discusses competence development strategies through the contribution of human resource development and training to increasing the company's profitability and how organizational change management through **portfolio project management** (Petrović et al., 2008) links modern human resource theory with the development of human resource organizational change management. The aim is to develop a model in the form of a researcher's contribution to creating a clearer picture of the key steps, metrics, and factors needed to understand how human resources and change management can impact corporate, national, and global development and growth (Djordjevic & Mihic, 2022), as well as to provide a deeper understanding of human resources as an important part of change management and lay the foundation for further theoretical and practical research.

The management of organizational changes, as a scientific field, includes various approaches, methods, and techniques that are appropriate for different business situations and the contingency prediction of factors influencing organizational changes. Numerous models of organizational change management in the literature differ according to their approach, with some advocating a top-down approach, others a bottom-up approach, and others still including a horizontal direction or a holistic approach to organizational change. Additionally, many authors outline different steps in organizational change management, which are mainly based on pioneering work in this area, such as Lewin's model, which consists of three phases: the unfreezing phase, the change phase, and the freezing phase. As a result, various steps can be found in the literature that simultaneously highlight key aspects of managing organizational change, such as awareness, leadership, **communication**, strategy, implementation, evaluation, and rewards (Dulanović and Jaško, 2008; Kotter, 2012; Cummings, 2010; Obradović et al., 2011; Tokarski et al., 2020).

The factors of change encountered in the literature are numerous, among which the following stand out: people, strategy, structure, and technology, with variations. The strategic analysis of change factors is particularly important, and different methodologies can be used for this, such as Internal Audits, SWOT, PESTLE, Balanced Scorecard, **Strategic Corporate Social Responsibility (CSR)**, and similar analyses (Coulter, 2010; Lauer, 2021; Jaško et al., 2013; Jovanović, 2005; Jovanović, 2008; Stouten et al., 2018; Djordjevic & Mihic, 2022; Mihic et al., 2019). These include both methodological and non-methodological approaches, as well as structural and descriptive representations.

This paper further investigates the differentiation of organizational change management factors from the perspective of people, particularly strategic human resource management, and explores the social phenomenon of engagement in relation to leadership and dynamic managerial competencies (Bradford & Cohen, 2017; Bradford & Burke, 2005; Collins, 2020; Davenport & Barrow, 2009; Petrović et al., 2008). It also examines the impact on motivation, abilities, creativity, and engagement of employees and teams through a human-oriented **organizational culture** (Fountaine et al., 2019; Djordjevic & Mihic, 2022), and from the third party to strategy (Ulrich & Grochowski, 2018), in the context of developmental teams and change leadership. This framework can help organizations achieve competitive advantages, improve organizational and business performance, and enhance dynamic organizational performance, which further contribute to increasing the level of human capital and the overall value of the company, also from corporate social responsibility (CSR) aspects and profitability (Bontis, 1998; Gupta & Roos, 2001; Mihic et al., 2019; Djordjevic & Mihic, 2022; Petrović et al., 2008). This includes intellectual, social, and human capital, leading to the successful management of organizational changes by realizing the full potential of people and the organization, as well as fostering creative business transformations. These transformations are driven by the creativity of employees, but also by the innovative ideas of stakeholders, which contribute to the overall innovation of the business - necessary for the survival, development, and growth of companies, from both small business incubator teams to teams of leaders of change in large corporations.

2 Literature Review

The management of organizational changes represents the organizational ability (Kotter, 2012) for constant change: structure, strategy, competencies, abilities (Dulanović & Jaško, 2008), market, management, employees, and processes (Anderson & Anderson, 2010). It is also part of the organizational development process (Cummings & Worley, 2009) and the innovation process, which consists of three phases: initiation, development, and implementation (Petrović et al., 2008).

Organizational change management, during the development of the scientific field, has resulted in a significant number of organizational change management models, with Kotter's model standing out as the most widely applied in practice. Additionally, when considering the factors, it is inevitable to take into account the effects of the forces of resistance to change and the forces instigating change (Kotter, 2012; Dulanović and Jaško, 2008; Petrović et al., 2008; Obradović et al., 2011; Marjanović, 2024), which are elaborated in this paper from the perspective of employees' resistance to change and their resilience in crisis situations, as well as from the perspective of internal and external drivers of change and the impact of strategic human resource management and leadership on the outcomes of organizational change management, both through case studies and theoretical viewpoints. Therefore, the development of employee competencies is one of the most important segments of effective organizational change management (Nielsen et al., 2010; Aguinis and Kraiger, 2009; Salas et al., 2008; Al Hamad et al., 2022; Dolan et al., 2022; Petrović et al., 2008).

2.1 The Importance of Strategic Human Resources Management for Change Management Process

According to Shuler (1992), strategic human resource management is the integration of business strategy, human resource policies, and practices across hierarchies, cultures, and values. It links, integrates, and coheres across levels in organizations, aligning with the strategic needs of the organization. Additionally, Boxall et al. (2000) suggest that the two dominant models of **HR strategies**, "best fit" and "best practices," are heavily influenced by national, sectoral, and organizational factors, as well as by strategic management and the "board's power" in determining its ability to perform its strategic role (Lengnick-Hall et al., 2009; Bozhinovska, 2019).

The development of human resources, as part of the emerging scientific field and practical area of organizational development - especially organizational culture - greatly affects the success of implementing organizational changes (Jaško et al., 2013; Ivanović, 2020; Flotman & Grobler 2020; Fenech, 2022; Khana and Ahammadd, 2020; Petrović et al., 2008). It also increases the level of readiness for change and the success of implementing organizational changes (Collins, 2020; Nurzaman et al., 2020). The role of change leaders, leadership abilities and skills, as well as the "personal concept of leadership" and dynamic managerial competencies, and "balancing the strategic and tactical approach" (Petrović et al., 2008) have a significant influence on the outcomes of managing organizational changes, especially in terms of changing organizational culture and behavior through human resource management strategies (Bradford & Cohen, 2017; Burke, 1998; Cummings, 2009; Petrović et al., 2008). Considering this, intellectual capital, especially **human capital**, which is an integral part of intellectual capital, contributes to the creation of a competitive advantage and the growth of the company's value. It creates opportunities for contributing to the wider social community and national and global growth and development through economic, legal, and ethical responsibilities (Lee & Hong, 2018; Salvatore, 2013; Janjić, 1998; Stanivuk, 2015; Mihic et al., 2019; Djordjevic & Mihic, 2022).

2.2 Organizational Change Model (Kotter) and HR Leader Model (Ulrich)

The models discussed in the case study research refer to Kotter's model (Kotter, 2012) of organizational change management and the Ulrich model (Scott-Jackson & Mayo, 2018) of human resource management, as the most adequate, based on numerous studies from relevant consulting agencies such as Deloitte, McKinsey, and Gartner. Their findings suggest that 80% of organizations manage change "from top to bottom," and most often use Kotter's model due to its clarity and simplicity in application. Additionally, considering the numerous authors in the emerging human resources development field, the Ulrich model is found to be appropriate for the field of change management research.

These models are used to provide a general overview of the case studies and analyze the relationship between organizational change management and the impact of human resource management in developing a new model of organizational change management. Kotter's model is one of the most well-known for applying change management principles in practical business examples. It belongs to the linear models of organizational change directed "from top to bottom," where the state of the organization and the behavior of employees change with the involvement of a change leader. Kotter's model consists of eight steps in organizational change management (Fig. 1):

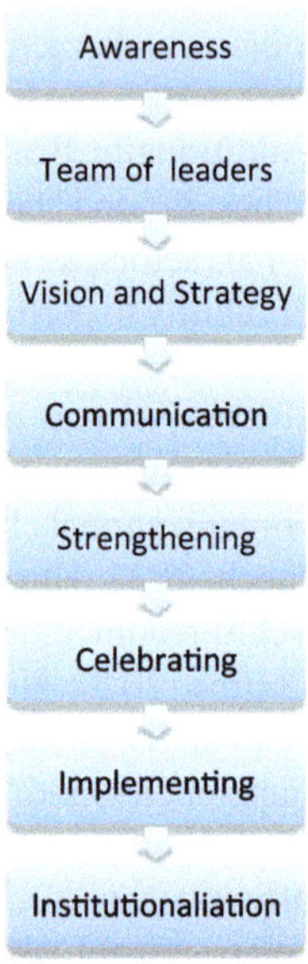

Fig. 1. Kotter's Eight-Step Model, Source: Kotter, P. (2012), Kotter J. P. (2014), Marjanovic M. (2015)

Developing Awareness of the Necessity of Change. The first step: Development of awareness regarding the necessity of change includes the following: analysis of the market, competition, and general environment; identification and analysis of key sources of potential changes; and creating a sense of urgency and encouraging action for change. Kotter's eight-step model begins with creating the preconditions for recognizing the necessity of change. Internal and external analysis of sources and factors of change contribute to the creation of precise indicators of business performance and the environment, which provide guidance to decision-makers. These indicators could be based on the portfolio project management concept, while simultaneously identifying possible crisis situations that the organization must prepare for in order to respond adequately to business challenges. Therefore, the model starts with spreading awareness and establishing a sense of urgency and necessity within the organization about the upcoming changes that need to be implemented, through specific indicators. It is also important to note that the model is based from the beginning on creating awareness among management and employees, which results in the need to announce and communicate the reasons and evidence for the change. This communication is essential to ensure that the change is understood and accepted, enabling the organization to maintain or reach the necessary level of business sustainability and ensure survival in the market. Additionally, this process helps ensure **business responsibility** and a wider contribution, both internally and externally, especially from a social perspective in these case studies. Therefore, the first step refers to spreading awareness about the necessity of change. This step is crucial because it creates the prerequisites for realizing the need for change, as well as the possibility of change, on the basis of which further effective management of organizational changes can be achieved.

Creating a Team of Change Leaders. Second step: Formation of the leading coalition includes the following: formation of an adequate team for change management and management of the change management group. This step refers to the establishment of the organizational change management mechanism, which could be defined by knowledge, competencies, skills, methods, and techniques. After awareness of the necessity of change has been established and accepted by decision-makers, it is necessary to configure a team of leaders who can implement these changes in a way that ensures the effective achievement of change goals. This also involves the further expansion of awareness, which is one of the most challenging tasks for today's leaders. This step represents a special challenge for the organization, as it involves configuring adequate teams of managers and change leaders who are aligned with business strategies, in this case, the goals of change. Change leaders, who are responsible for managing organizational changes, must possess adequate knowledge, abilities, skills, and special characteristics relevant to business as well as to leadership, ensuring that changes are managed effectively. Therefore, in addition to managerial knowledge and abilities, it is important to consider the potential of management skills when configuring the team. The leader's characteristics, leadership styles, and the potential influence of the leader are crucial. Furthermore, the team of change leaders must be configured in such a way as to provide the necessary types of authority. Both formal and informal authority are of great importance in gaining the necessary level of trust from employees in the change leaders. This step is a challenge for the organization, as the initial step must ensure the expansion of awareness among decision-makers, while simultaneously expanding the awareness of potential change leaders, guiding them in the desired direction of change. This ensures that leaders with the necessary competencies and potential influence are engaged in the team responsible for managing organizational changes, contributing to the further expansion of employees' awareness of the necessity of implementing changes and increasing the strength of the change movement, as well as developing employee resilience.

Spreading the Vision and Strategy of Change. The third step includes the following: determining a clear vision and change strategy, creating a vision, and developing a strategy. This step focuses on developing a vision and strategy to guide change. After forming a team of leaders who will lead the changes, it is necessary to create a vision of change that effectively engages cognitive competencies and forms a set of opinions in a way that ensures the understanding and trust of everyone in the organization regarding the goals and outcomes of the changes. The challenge of this stage is rooted in human nature, which often leads to a negative attitude and mistrust when facing change. Therefore, in this step, in addition to creating a vision, a change strategy is also developed to clearly and comprehensively outline the expected positive outcomes of the changes. This helps engage employees in the activities of implementing changes and reduces their resistance to change. In this context, spreading awareness and creating a vision and strategy can be considered inseparable steps of change management. In alignment with changing beliefs and attitudes, based on facts and evidence of the necessity of change, it is essential to design and create a vision and strategy for the organization's direction, with clear strategic goals and action plans that can guide the organization in its future operations. In addition to considering the impact on employees and ensuring the vision and strategy are aligned with their needs, it is important that the vision is based on external and internal analyses of change factors. This will enable the creation of a strategy grounded in relevant data, facts, and predictions of change factors.

Communication. The fourth step includes the following: communicating the vision and strategy of change, using various methods, techniques, tools, and communication channels to spread the vision and strategy, and representing the leading coalition of leaders as a model for change. This step, which follows all the previous steps, focuses on communication. In this phase of change management, it involves the creation and implementation of communication strategies for the previously developed vision and change strategies, through the team of change leaders. In this step, change leaders are committed to continuous communication with employees in order to enhance understanding of organizational changes and their outcomes. Through the strategy of leading by example, change leaders can become leaders of new sets of opinions, beliefs, attitudes, and assumptions, which form new value systems and organizational behaviors. These new behaviors change the actions of employees by demonstrating the behaviors expected from them, in line with the envisioned changes. Communication can be formal, informal, direct, or indirect, through examples and various other methods, techniques, and communication channels that can be used in this step. The goal is to convey the vision of change and create new systems of values and behaviors that align with the goals of the change, which contribute to transforming business processes, structures, strategies, technologies, and more.

Strengthening Actions. The fifth step includes the following: involving employees in the change process, removing obstacles to change, modifying systems and structures that hinder the implementation of the vision for change, and encouraging a proactive approach to change. During the implementation of changes, the fifth step involves activities related to aligning the elements of the organization with the new vision and business strategy. Changes in organizational structure, technology, business processes, strategies of organizational units, systems, etc., can be of strategic importance, and most often are. To implement the new vision and strategy, it is necessary to adapt all relevant elements of the organization to optimize the process of implementing organizational changes. At the same time, it is essential to empower employees for change; in addition to increasing their resilience, it is also necessary to motivate them and develop their creative potential in order to engage them and contribute to the innovation of the organization and the successful implementation of changes.

Celebrating Minor Team Victories. The sixth step includes the following: creating and highlighting short-term successes, planning exact performance improvements, creating successes, recognizing and rewarding achievements. According to the model, the sixth step is marking the achievement of short-term goals, ie. celebration of the first visible results and all subsequent short-term results until the realization of the long-term goal. The implementation of this step contributes to the creation of a positive attitude about changes, providing additional motivation to continue the implementation of changes. Exact indicators of business performance can be used to present the successes that have been achieved, and based on the results, reward systems can be formed and recognition awarded.

Implementing Further Changes and Final Imprinting. The seventh step includes the following: continuing the implementation of changes, increasing the level of credibility in support of further necessary changes, developing human resources with the potential to realize the vision of change, and introducing new developmental changes. The seventh step of the model refers to the persistence of management, the change leadership team, and employees in the continued implementation of changes. This is the step in which new ways of doing business and the new direction are integrated into the organization. After evaluating short-term business performance and establishing an adequate reward system, feedback and corrective action activities are implemented in this step. Based on the further expansion of awareness regarding what has been achieved, the vision and strategy of changes are revised and incorporated into new strategic plans and development programs, especially the human resources development plan, which is based on the potential for further change implementation, increasing credibility, and introducing new developmental changes.

Institutionalization of Changes. The eighth step includes the following: institutionalization of the achieved changes, increasing the quality of leadership, management, strategy, and orientation, highlighting the clear correlation between new forms of behavior and business success and performance, and the development of change management systems and change leaders. The final step of Kotter's model refers to the institutionalization of the achieved changes. Therefore, based on the revision of the vision, strategy, and development plans according to the results achieved in the previous step, this step focuses on long-term planning to increase the quality level of future visions and strategies, using specific correlation indicators for the changes implemented and dynamic organizational performance. At this stage, it is essential to plan the further development of change leaders to ensure sustainable change management.

Mistakes in Change Management Kotter also points out the following common mistakes in change management:

Excessive indulgence in the satisfaction of achievements
Formation of insufficiently powerful leading coalition
Underestimating the power of vision
Not good enough communication
Allowing obstacles to obstruct vision
Inadequate marking and recognition of short-term successes
Premature announcement of victories
Changes that are not firmly embedded in the ***organizational culture***

According to authors William Scott-Jackson and Andrew Mayo (Henley Business School and Oxford Strategic Consulting, 2018), Dave Ulrich created a new perspective on the human resource function in organizations. The human resource management model he developed views human resources from a central leadership position and assigns the role of change leader to human resources. The dimensions, when converted into roles, can be described as follows (Ulrich, 1997; Scott-Jackson and Mayo, 2018):

The First Dimension and the Role of Human Resources. Strategic Partner - Management of Strategic Human Resources: Aligning HR and business strategy, executing strategy, human resource planning, knowledge of business, change, consulting, and learning, partnering with line managers, strategy formulation, and change implementation. The strategic partnership between human resources and business leaders involves creating a business strategy based on the consultative role of human resources in business, as well as strategic planning of human resources through internal and external analysis of the business environment, and the application of various methods and techniques of strategic planning. Experts and professionals in the field of human resources should possess knowledge of business, change, consulting, and learning and development. Additionally, in the role of a strategic partner, human resources manage the implementation of organizational changes in partnership with line managers.

The Second Dimension and Role of Human Resources. Administrative Expert - Management of Infrastructure: Managing processes and services, building efficiency, functional expertise, specialist knowledge, HR practice, and organizational effectiveness. Apart from the strategic partnership, the human resources model developed by Ulrich retains a functional role, which is fundamental to the management of processes and services carried out by the human resources management function. In this role, human resources should possess expert knowledge in the field of human resource management within the organization, as well as specialist knowledge and practices. Additionally, they should have an understanding of organizational effectiveness and the construction and development of operational processes in human resources, to achieve organizational effectiveness and efficiency.

The Third Dimension and the Role of Human Resources. Listening and responding to employees, increasing commitment and capability, employee protection / Employee Advocate, managing work conditions and well-being, listening to employees' opinions and needs, and managing diversity. According to the model, human resource management, in addition to its strategic, operational, and functional roles, involves planning, organizing, and controlling human resources, while also addressing the social-psychological aspects of business, with a strategic and operational orientation toward human resources. The interaction between the employer and employees is created and implemented through the role of human resources in the organization, which focuses on employees as individuals, not just processes. The strategy is designed to increase employee contribution by actively listening to and responding to employees, enhancing their commitment and abilities, ensuring their protection, and safeguarding their physical and psychological well-being. Additionally, this role ensures fair and equal interpersonal relationships, including those with the employer, as well as managing work conditions and employee well-being. All of this is done in alignment with available opportunities, needs, and through active listening and valuing employees' opinions, attitudes, and needs, with the ultimate goal of creating strategies and business plans that can increase employee contribution.

The Fourth Dimension and the Role of Human Resources. Focus on staff as people, learning, development, effective teams, coaching, talent management. Human capital development represents another dimension of Ulrich's human resources model. It focuses on employees as individuals, so, like the previous dimension, the emphasis is not on human resource processes but on people and their inherent nature. This dimension aims to develop the full potential of employees, i.e., their talents, starting with social and emotional intelligence, which are crucial for unlocking employees' creativity. It also aims to foster the development of knowledge, abilities, skills, personal characteristics, motivation, satisfaction, and creativity - elements that represent the potential of talents within organizations. This potential contributes to achieving business goals, driving business innovation, and allowing employees to achieve self-actualization. The development of human capital applies not only to employees but also to line managers and business leaders.

The Fifth Dimension and the Role of Human Resources. Management of transformation and change, ensuring capacity for change, creating a renewed organization, HR leaders, functional leadership. In line with the development of human capital, it is essential to develop leaders, experts, and specialists in the field of human resources. This dimension of Ulrich's model focuses on the leadership potential of human resources in the context of change management, positioning HR as a change agent responsible for transformation and change management. Through the partnership between functional leaders and human resource management leaders, business is viewed from the perspective of innovation potential, with a focus on the necessary management of organizational changes. This dimension places human resources in the role of leaders in organizational change management (Fig. 2).

Thus, the human resource management model developed by Ulrich (1997) views human resources from the central position of leadership and assigns the role of change leader to human resources, based on three pillars: the first pillar, traditional functionalities, such as resourcing, payroll, monitoring, and employee relations advice; the second pillar, the center of excellence, such as expert teams and development projects in areas like reward, learning, engagement, and talent management; and the third pillar, strategic partners, where HR acts as a partner to business leaders in influencing strategy, monitoring its implementation, and optimizing the use of human capital, as well as fostering the development and self-actualization of human capital (Scott-Jackson and Mayo, 2018).

However, some of the shortcomings of the model and the challenges faced by human resources in the form of a partnership lie in the organizational structure, which does not clearly define and configure the partnership within traditional organizational structures and schemes. This issue can be more complex in the case of matrix and network structures, where it may be a challenge to realize the full potential of the partnership role of human resources. Another shortcoming of the model is the relationship between line managers and the partnership role of human resources, which often remains established or tends more toward an advisory role, with the partnership role still evolving (Scott-Jackson and Mayo, 2018).

Fig. 2. Ulrich's HR Leader Model, Source: adapted from (Ulrich, 1997; Scott-Jackson and Mayo, 2018)

3 Methodology

This paper presents a case study (Marcyik et al., 2005) as an example of company transformations in the field of mergers and acquisitions, and analyzes the transformation process in the supply chain of a company operating in Central and Southeast Europe. The data collected includes both primary and secondary data. Secondary data were gathered from consulting materials and literature through a search on the Google Scholar platform, scientific research articles and journals, as well as books in their entirety or in available parts and chapters. The primary data collected for this study refers to archival materials in the field of human resource management, based on the professional experience of the author, as well as data collection and processing techniques using a semi-structured interview. A semi-structured interview was conducted in July 2024, lasting between 45 to 60 min. The interview focused on roughly defined areas of questions related to organizational change management models, human resource management models, and the mutual influence of these two areas and organizational change management factors, in line with the subject and objectives of the research, along with additional questions. The model and process of organizational change management are analyzed from the perspective of a change leader in two case studies (Table 1).

Based on the collected scientific literature and a semi-structured interview related to the subject and objectives of this research, a case study analysis of supply chain acquisitions was conducted, which is presented in the following sections of this paper.

Table 1. Fields of semi-structured interview

Number of Field	Field
I	Organizational Change Leadership
II	Human Resource Management Strategies

4 Case Study

4.1 Organizational Transformations and Human Resources

Organizational transformations can be categorized into various types. These case studies focus on organizational change and the impact of human resources strategies, specifically within the category of organizational transformation known as mergers and acquisitions (M&A). These types of organizational transformations in the supply chain are also strategic moves aimed at securing competitive advantages and ensuring long-term sustainability, as well as corporate social responsibility, in a challenging economic, ethical, and legal business environment (Mihic et al., 2019).

The global market for mergers and acquisitions in supply chains experiences a high level of volatility, driving the conquest of new markets, growth, development, and increased product availability to the population. The market in Central and Southeastern Europe stands out in terms of market entry and the expansion of large multinational corporations through mergers and acquisitions of smaller companies. Globally, according to data from the Institute for Mergers, Acquisitions, and Alliances (2024), the volume of mergers and acquisitions reached approximately 50,000 per year from 2015 to 2019, with more than 790,000 transformations completed since 2000, totaling $57 trillion USD. Furthermore, geopolitical tensions and digital transformations contribute to an increase in supply chain deals (Supply Chain Digital, 2024). According to Ernst and Young Global Limited (2024), the Central and Eastern Europe (CEE) region recorded 224 deals worth $54 billion USD in 2024, compared to a 58% decline in deal value and a 27% decline in deal count compared to 2023. According to the Global M&A Deals in Q2 2024 report (2024), supply chain deals top the list with 200 deals valued at $39 billion USD across 19 transactions. This strategic focus on the supply chain reflects companies' increasing recognition of the need to enhance resilience, agility, and efficiency within their operations.

The development of the field of human resources management has contributed to the emergence of new approaches, with the direction of human resources development being singled out as a separate part of strategic human resources management. In recent years, companies have increasingly focused on the development of human resources, with more organizations implementing these concepts through strategic direction and organizational structure (Petrović et al., 2008). Digitalization processes, as a priority for the Republic of Serbia, have contributed to exports nearing EUR 1 billion (Djordjevic & Mihic, 2022). Additionally, the development of technology has led to major changes in information technology, with entirely new high-tech industries emerging and existing ones undergoing transformation (Petrović et al., 2008).

Given these developments, the introduction of new jobs and the creation of entirely new workplaces require, in addition to the necessary development of human resources for the evolving needs of the labor market, the development of new management models (Petrović et al., 2008). This also includes the management of organizational changes to ensure the adequate management and implementation of these changes.

With the development of new forms and types of work engagement, new forms and types of organizational cultures are also emerging (Petrović et al., 2008; Djordjevic & Mihic, 2022). These developments can, to a certain extent, establish differentiated motivational factors and factors that influence employee satisfaction levels, particularly with the entry of "Generation Z" into the labor market, following the "Millennial" generation.

In response to different interests arising from the development and increasing use of new technologies in both everyday activities and business, management styles are also evolving - particularly in the context of change leaders and managerial competencies, which are aligned with the transformative processes occurring within companies.

The labor market in recent years reflects these changes, starting from crisis situations like the Covid-19 pandemic to the ongoing challenging geopolitical issues facing humanity. These factors impact both job supply and demand, creating unique flows of human capital in the global labor market and presenting challenging indicators for future trends in global human capital movement. This, in turn, influences strategic directions, particularly in the area of corporate social responsibility (Mihic et al., 2019).

Considering the natural demographic changes of the population from the aspect of the hourglass phenomenon, the current indicators point to a demographic structure represented by an inverted pyramid, where the elderly population is at the top, and the younger population forms the narrowest part. The ideal social structure is described as a regular pyramid, where the widest part would refer to the youngest population of society, with the oldest population at the very top of the pyramid.

Observing the demographic indicators, it can be established, based on statistical reports from the Eurostat portal (2024), that the population is aging, which further results in a decrease in the total number of inhabitants in Europe, given the lower birth rate compared to the mortality rate. Considering the causes of the negative gap between birth rates and mortality rates over the last 20 years, some of the reasons that can be cited include crisis situations that humanity has been exposed to, starting with the 2008 World Economic Crisis, followed by more recent crises that are still ongoing, and the trend of an aging population.

To explain in more detail, natural population changes can be indirectly related to crisis situations and times, which negatively affect people's living standards, economic and psychological security, and, therefore, the labor market and labor force. In particular, the current trends in demographic factors show the aging of the population, which predicts a higher proportion of older people compared to younger people. This, in turn, impacts strategic business decisions, especially regarding the types, conditions, and economic factors of jobs that will be needed in the coming years, as well as the labor market and labor force.

Migration is also an established phenomenon, driven both by economic reasons and other factors, primarily related to the satisfaction of existential needs. Between 2022 and 2023, there was a trend of growth in hourly labor costs in the European Union by

5.3%, due to strategic changes in the field of work and the workforce. By 2050, it is expected that 80% of elderly people (the population over 60 years of age) will be in less developed and weaker countries, according to statistical reports from the World Health Organization portal (2024). Additionally, according to the same portal, in 2020, a larger share of the population was recorded in older age groups compared to the share of younger populations (under 5 years old). Therefore, it is not surprising that this trend will continue. By 2050, the share of elderly people in the total population is expected to double, from 12% to 22%.

The presented indicators are also influenced by the trend of increasing the age of the population. It is predicted that by 2050, the share of the oldest population (people over 80 years old) will triple, reaching 426 million people, while the number of people over 60 years old will rise to 2.1 billion. At the same time, a similar inverted hourglass phenomenon has been observed in the United States of America, with a disproportionately larger central part of the pyramid compared to the top and bottom. This explains the baby boomer generation, which, in recent years, has moved toward the top, thus establishing the hourglass phenomenon. In this scenario, the most numerous population (baby boomers) is at the top of the pyramid, while the next most numerous population (Generation Y) is at the bottom. The smallest population (Generation X) is in the middle.

The projection of the demographic structure of the population in the Republic of Serbia between 2030 and 2040, based on absolute data and monitoring the trends from the previous ten years, as provided by the Republic Institute for Statistics of Serbia (2024), is shown in Fig. 3. The average age of the population in the Republic of Serbia is 43.5 years for both women and men. The population aging index (the ratio of the elderly population aged 60 and above to the young population up to 19 years) is 144.5%, which is the highest value recorded in the last ten years. Furthermore, the growth of the population aging index has been consistently increasing.

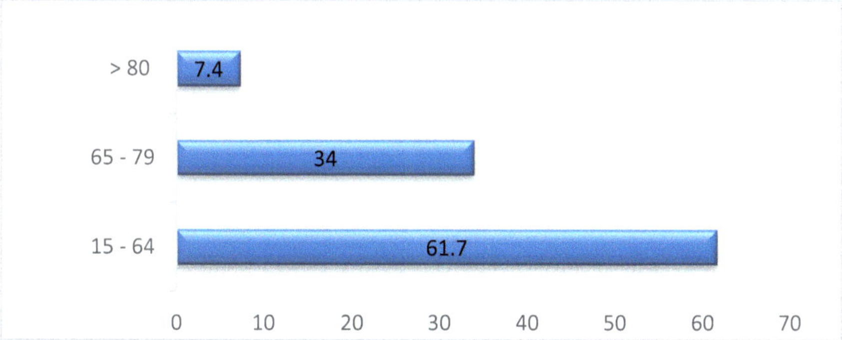

Fig. 3. Projection of the demographic structure of the population in the Republic of Serbia in the period between 2030 and 2040

In the labor market of the Republic of Serbia, the unemployment rate has consistently declined over the past ten years. In the last quarter of 2023, it stood at 9.1%, according to data from the Republic Institute of Statistics of Serbia (2024). This is an important economic indicator that reflects constant growth, including the increase in average salaries and the growth in registered employment. These factors contribute to maintaining economic stability and enhancing the economic security of the workforce and human capital. When examining the unemployment rate by education level in the Republic of Serbia, a gap of 44.4% can be observed, which corresponds to the unemployment rate of the medium-skilled workforce. This gap further presents opportunities for additional qualifications in line with the upcoming changes in transformative processes and human resources management models for change and development.

When considering migration and the flow of human capital, one can observe, based on statistical reports from the International Labour Organization (2024), the geographical distribution of the global labor force, as well as the distribution of this labor force by age. The largest share of the global labor force is located in Asia and the Pacific, which accounts for more than half of the total labor force, while only 9.9% of the global labor force is in Europe. Specifically, 6% is in Northern, Southern, and Western Europe, and 3.9% is in Eastern Europe. Additionally, when examining the employment engagement of the younger population, specifically Generation Z and Millennials, a notable gap appears in comparison to the older populations of Generation X and Y. Globally, the youth unemployment rate is 13.3%, while the unemployment rate for adults is 3.9%. A particularly interesting indicator, considering the subject and objectives of this research, is the motivation strategies specifically targeting the younger generations, such as Generation Z, who are entering the labor market. This trend can also be observed when comparing developed and underdeveloped countries. In highly developed countries, the youth unemployment rate is 10.4%, while the adult unemployment rate is 3.8%. In less developed countries, the youth unemployment rate is 9.2%, and the adult unemployment rate is 4.5% (Fig. 4).

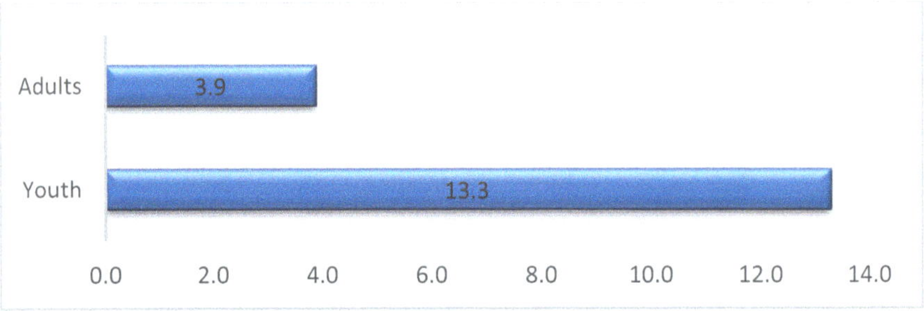

Fig. 4. Global unemployment rate, comparative view of youth vs. adults (%)

Also, according to the statistical data, there is a significant deviation in earnings within the group of information and communication activities. This group, in addition to publishing, program activities, and broadcasting, also includes computer programming, consulting services, and information service activities. These sectors represent the fastest-growing part of the gig economy, which primarily involves the younger labor force population between the ages of 15 and 30.

According to the graphs and data presented in the previous section of this research, a gap is evident between the skills and abilities present in the labor market and the needs of employers. Additionally, the rapid development of the technological sector, which continuously leads the market in terms of job offers and working conditions, can be observed. This sector also generates strong potential for growth and expansion (Djordjevic & Mihic, 2022).

The changes in the labor market, driven by transformative processes such as digitization, mergers, and acquisitions, allow for a closer examination of the strategic aspects of change management in human resources. Specifically, changes in the external environment create the necessity for transformations within companies themselves, leading to the emergence of completely new jobs, workplaces, professions, motivational factors, and organizational cultures. This also includes changes in educational systems, decision-making processes, and strategies based on corporate social responsibility (CSR) that can enhance the quality of employment, promote fairness in the workplace, ensure the physical and psychological safety of employees, and support individual and professional development. These factors contribute to overall economic and social security, stability, and growth, which in turn have a positive effect on reducing poverty and inequality.

These case studies focus on organizational change and the impact of human resources strategies, specifically within the context of organizational transformation, such as mergers and acquisitions (M&A), in Central and Eastern Europe.

5 Discussion

5.1 Organizational Change Process

In the examined cases, as well as in the literature on mergers and acquisitions, the process of managing organizational change can be viewed in two phases. The first phase involves the preparation and decision-making regarding the acquisition as a transaction, while the second phase represents the post-acquisition process, which focuses on the integration of two or more organizations during which organizational change is implemented. The first phase includes the decision to initiate organizational change and the transaction between the companies, while the second phase involves implementing organizational changes and integrating two organizational systems.

From the perspective of organizational change management, both phases are equally important. However, the second phase - the implementation phase, or the post-acquisition period - is often considered more critical in terms of the effectiveness of change management, and thus the overall success of the acquisition. This phase requires the achievement of clearly defined business goals and strategies, which largely depend on managerial and leadership competencies in the area of change management, as well as on the methods and techniques used.

The realization of the acquisition transaction can also be considered from the aspect of organizational change management. It involves preparing for the acquisition based on an analysis of both external and internal factors driving organizational changes, strategic analyses, plans, financial assessments, and evaluating the costs and benefits of new acquisitions. This includes considering the profitability of investments, the investment return period, income from entering new markets, the expansion of the distribution network, and sales channels, relative to total investments and new market potentials. Therefore, the acquisition of new branches can be realized by creating a feasibility study, often through portfolio project management and a detailed organizational change implementation program (Jovanović, 2005; Petrović et al., 2008).

A decision based on long-term strategic planning, internal and external analysis, and consulting support plays a significant role in the change drivers, such as the owners and supervisory boards of the corporation, investment groups, or conglomerates. This phase of organizational change should be considered as part of the organization's development and business growth in line with portfolio project management and general management principles. Thus, the first phase in this case study can be viewed as the strategic management phase, where a formal acquisition decision is made.

On the other hand, this process can also be considered in three phases, according to Petrović et al. (2008). The initial phase in managing organizational changes is where external and internal analysis provides insight into the drivers of organizational change, essentially expanding the organizational awareness. During the strategic analysis, an important dimension is intellectual capital, particularly in terms of human resources planning. This includes the availability of qualified labor, necessary training and development of the workforce, as well as trends in the labor market, which should be projected in the feasibility study for the expansion of the supply chain market and portfolio project management program.

At the same time, the analysis of the labor market and workforce can be strategically aligned with the CSR (Corporate Social Responsibility) concept, with an increasing emphasis on the use of local labor (Mihic et al., 2019). This aligns with expanding the availability of goods and services, and creating opportunities for the development and growth of both local and global communities (Fig. 5).

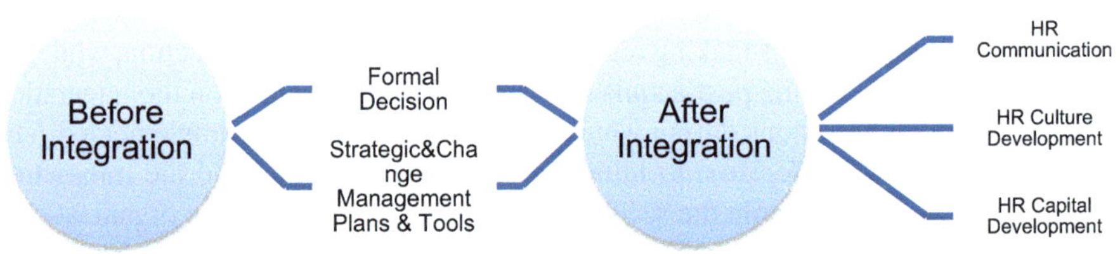

Fig. 5. Phases of organizational change management

A large number of studies highlight the significantly high failure rate of change management initiatives, which can be as high as 70%. From the perspective of the organizational change management model, particularly in the field of mergers and acquisitions, the Lean IX M&A Survey (2021) identifies the post-integration period as having the greatest impact on the outcomes of organizational change management, accounting for 89%. In comparison, due diligence has an impact of 44% on the outcomes of organizational change management.

According to research conducted by the management consultancy Gartner (Changing Point, 2024), the average organization had five change initiatives over the past three years, and 73% of organizations expect an increase in the number of change initiatives in the coming years. Nearly 100% of companies consider change management to be either an important or extremely important task for the future. Furthermore, more than 200 human resource management experts have identified "Change Process Monitoring" as the most critical tool for developing the workforce of the future, followed by coaching (Lauer, 2021).

5.2 Organizational Change Factors

In this case study, one of the most important factors in organizational change management during acquisitions is human capital. This factor plays a crucial role not only in the decision-making process for acquisitions but also in general business development and in achieving the effectiveness of organizational change.

The trend of diversification, not only in terms of products and services but also in strategic business areas such as investments and acquisitions in entirely different sectors (e.g., the IT industry), represents a key strategic factor in managing organizational change. This diversification is closely linked to the human aspect of organizational change from the perspective of human resource management. It also impacts the growth of the workforce and talents, as well as the development of human capital, contributing to both local and global growth. Based on strategic analyses of trends and current market conditions, business development strategies are crafted, awareness of the necessity of change is spread, and teams of leaders are created to manage organizational changes.

According to Djordjevic and Mihic (2022), the investment rate in the IT sector exceeds 80 EUR per capita, with a growth trend that is expected to reach 150 EUR by 2025. Although this is below the EU average of 800 EUR, it highlights the importance of digitalization. The development level of the IT industry is so high that, by 2019, 337 software companies were founded in Serbia, and 8 out of 10 of the leading Serbian software companies have foreign ownership (Djordjevic & Mihic, 2022). The Government of Serbia recognized the need for change and directed efforts through special programs and benefits in this area. As a result, a completely new industry branch was developed, which not only reduced the unemployment rate but also contributed to unlocking the potential of human capital, positioning Serbia on the global IT sector map and fostering acquisitions in the sector (Djordjevic & Mihic, 2022).

According to international research by the International Institute for the Organization of Learning and Innovation at the University of St. Gallen, it was proven that in 111 companies in the German-speaking region, successful changes were linked to certain relative factors, reducing costs by 25% and required time by 16% (Lauer, 2021).

5.3 Organizational Change Strategy

In this part of the case study analysis, we will discuss the applied organizational change management strategies in conjunction with human resource management strategies, supported by available literature and descriptive statistics.

According to data from a panel discussion at the International Symposium (Symorg, 2024), change management strategies play a crucial role in the successful implementation of organizational changes. A key point emphasized by the audience was the lack of an adequate change management strategy in the current transformative processes driven by the digitalization of business. According to Gartner's research through the Changing Point platform (2024), modern organizations with flexible structures and complex reporting lines ensure the success of change management by implementing open-source change management strategies. This approach increases the likelihood of success by 22%, reduces implementation time by one-third, and decreases the total time required for change by 12 h per week per employee. Furthermore, the success of change efforts increases by 24% when employees are actively involved in the implementation process.

Human Resources Strategies in Change Mangement. Human resource management strategies have a positive impact on productivity growth by enhancing the effectiveness of intellectual capital, strengthening employee adaptability and resilience, and improving financial performance. These strategies help reduce labor turnover and increase profit per employee (Coulter, 2010; Guest, 2017; Makki & Lodhi, 2014; Ognjanović et al., 2017; Dulanović & Jaško, 2008; Mihic et al., 2019). Recruitment, development, and training of employees, as part of strategic human resource management, contribute to the formation of human capital and the effectiveness of leadership in managing organizational change. With an adequate analysis of the labor market and workforce, it becomes possible to design strategies for developing intellectual capital, which in turn support business development and organizational change management strategies (Antonelli et al., 2010; Sitar & Vasic, 2004; Ognjanović, 2016; Nawaz, 2018; Asikhia et al., 2021; Collins, 2020; Dahiyat et al., 2020; Dragicevic & Mihic, 2020).

From the perspective of strategic human resource management, several studies highlight that an increase in employee autonomy has a significant positive effect on employee well-being (Hewagama et al., 2019). Conversely, an increase in the level of fairness has only a partial and mediating effect on well-being (Zhang et al., 2020). Employee productivity is positively influenced by training and the development of employee motivation (Ajmal et al., 2016), with job satisfaction being a key driver of motivation (Khan, 2012; Nurlaila, 2022). Job satisfaction is further influenced by the intention to stay longer in the same workplace (Snape & Redman, 2003) and the development of emotional intelligence (Sabie et al., 2020). Additionally, satisfaction with leadership strategies plays a significant role in enhancing job satisfaction, and increases in wages and benefits have the greatest impact (Kinnie et al., 2005).

HR communications and employer branding have a strong impact on employee performance, particularly through activities that develop awareness and align individual and organizational goals. These activities contribute to increased employee satisfaction, retention, loyalty, engagement, well-being, and the development of agility and resilience (Lievens & Slaughter, 2016; Fernandez-Lores et al., 2016; Caputo et al., 2023; Monteiro et al., 2020; Eger et al., 2019; Alniacik et al., 1998; Hanelta et al., 2021; Kaur Bagga et al., 2023; Khana et al., 2020; Barjaktarović et al., 2021).

Strategies Based on External and Internal Analysis. Furthermore, in these case studies, an analysis of the labor market for qualified labor in the relevant supply channels reveals a deficit in the workforce within the industries where the companies being acquired operate. Expanding the market and entering new regions within the same industry thus requires specialized human resources management strategies. These strategies should focus on the development, retention, and well-being of employees to minimize labor turnover in the integrated companies, as well as on local efforts by employers to empower new talent. The core human resource management strategies used in the case studies emphasize internal communications, which are deployed by change leaders to foster increased trust among employees in the company's management. This is a critical aspect of the second phase of the organizational change process–the post-integration phase–where the success of the merger or acquisition largely depends on the effectiveness of the organizational change management. Effective change management strategies in this phase are essential for the long-term success of the integration. Additionally, internal and external audits are integral components of portfolio project management. These audits help identify potential unpredictable challenges and assess other business areas that may be difficult or impossible to predict (Djordjevic & Mihic, 2022; Mihic et al., 2019).

Strategies Based on Change Leadership Development. Therefore, the second phase involves the influence of human resource management, with the joint responsibilities of mergers and acquisitions (M&A) managers and human resources (HR) managers. In this stage of organizational change management, the roles of M&A and HR managers overlap to some extent, requiring a high level of collaboration. This collaboration demands a broad range of knowledge, managerial competencies, skills, and leadership styles, which can be ensured through adequate development and training before the acquisition. Additionally, it is essential to reduce the effects of "organizational silos" to achieve effective coordination in managing organizational changes and realizing the potential of leaders, ultimately leading to the successful management of organizational transformations. Line managers also play a crucial role in this process, particularly in newly acquired companies. Thus, the dual role of line managers, M&A managers, and HR managers requires knowledge and skills in both human resource management and leadership. This ensures the effective configuration of the leadership team responsible for managing organizational changes. Moreover, the importance of tacit knowledge, "soft skills," and personal characteristics, as well as authentic leadership styles, is emphasized. These factors contribute to a greater influence on employees in integrated companies during the implementation of organizational changes. In addition to the dual role of managers, leaders emphasize the need for multidimensional knowledge, including expertise from various relevant business areas. Therefore, the leadership team for managing organizational change should comprise experts in M&A, HR, and supply chain management. Furthermore, the change management leader must possess specific knowledge in these areas to effectively guide the change process. To simplify, it is clear that the leader must have both technical and managerial expertise to gain credibility from employees and build trust during the acquisition process.

Through networking, consulting, and skill expansion, not only is human capital developed, but entire industries or sectors are also enhanced. Managers are guided toward developing leadership skills, as well as knowledge in tactical and strategic management methods and techniques, in order to achieve growth in return on invested capital. Research indicates that while technical knowledge of a leader has only a 10% impact on management factors and outcomes, strategic knowledge contributes 50%, and managerial knowledge contributes 40% (Petrović et al., 2008).

Numerous studies emphasize the importance of managerial competencies in change management. For instance, research reveals that 31% of general directors are dismissed due to poor change management. Furthermore, only 26% of organizations have leaders with adequate change management abilities and skills. A major global study, DDI Global

Leadership Forecast 2021, conducted by a consulting firm, included 2,102 HR experts and 15,787 leaders across 50 countries and 24 industries, highlighted the development of new generations of leaders, particularly employees between the ages of 21 and 38. In comparison to current leaders, who score 53% for communication and interaction skills and 48% for empathy and emotional intelligence, the new generation scores 45% and 39%, respectively. HR managers, when asked about priorities for the next decade, emphasized the importance of the following areas: employee development (84%), flexible working arrangements (75%), workplace diversification (72%), leadership aspiration (72%), and female leadership (59%).

Research from Changing Point (2024) reveals that 37% of employees resist change, with key reasons including a lack of trust (41%), lack of clarity about the need for change (39%), fear of the unknown (38%), changes in job roles (27%), and exclusion from the decision-making process (23%). Additionally, 28% of employees believe that top-down change management increases anxiety and anger, contributing to resistance. According to the CEB Corporate Leadership Council (2024), employee responses during changes include: 5% expressing anger, 29% expressing anxiety, 35% expressing hope, and 59% expressing pride. However, 73% of employees report feeling high stress during changes, 40% feel anxious, 28% feel angry, and only 16% feel happy. Moreover, 83% of employees believe their employers lack the necessary tools to adapt to changes.

Strategies Based on Human Capital Value. An essential factor in successful change management is establishing a sense of trust and security for employees during and after integration, particularly in industries facing a labor shortage, as is the case in this study. Therefore, change management strategies are implemented alongside human resource management strategies, with the primary goals of retaining and developing talent, while enhancing employee security, satisfaction, and motivation. These strategies also aim to reduce employee turnover costs, particularly in a labor market with a shortage of qualified workers, which in turn strengthens human capital.

The adjustment of integrated human capital through targeted training and the direct support of change leaders contributes to achieving previously defined goals such as work quality, productivity, and employee commitment. Adequate training and development help bridge knowledge gaps and align skills and managerial competencies across the integrated companies. After the acquisition decision is finalized, trained change leadership teams focus on spreading awareness of the need for change, while promoting the vision and strategy for the transition. These efforts are primarily aimed at increasing employees' sense of security, satisfaction, and motivation, which in turn helps reduce turnover and resistance, while enhancing resilience to the crises often associated with organizational change.

These strategies also facilitate the synergistic effects of acquisitions by increasing the value of human capital. In this context, organizational change management serves as the intersection between corporate objectives and individual goals within the corporation.

Corporate Social Responsibility (CSR) strategies are also crucial in enhancing efforts to source new talent and increase human capital value (Mihic et al., 2019). Additionally, business incubators, which drive technological innovation and empower the workforce (Djordjevic & Mihic, 2022), and the rise of new work engagement models, such as the GIG economy and the IT sector, contribute significantly to business development

(Petrović et al., 2008). The IT sector in Serbia, for instance, has become a leader in the global IT industry, with several acquisitions of established companies, many of which began as business incubators and are now integrated by global IT giants (Djordjevic & Mihic, 2022).

Many scholars highlight the critical role of intellectual capital in strategic management, particularly in motivating direct foreign investments, mergers, or acquisitions. Intellectual capital is closely tied to "exclusive knowledge" and "management skills" and contributes to the synergistic effect that enhances a company's value (Salvatore, 2014; Gupta & Roos, 2001; Kalenyuk et al., 2020). Human capital plays a significant role in fostering innovation, business performance, and sustainable growth, especially in leveraging its full potential (Ali et al., 2021; Bontis et al., 2004; Kovjanić & Vukadinović, 2021; Krstić, 2009).

In the case studies analyzed in this paper, human capital is particularly crucial during both the post-integration period and the pre-transaction phase, as these stages represent opportunities to harness the potential synergies of human capital integration. Bridging the gap between existing and desired levels of employee knowledge, skills, and abilities is key to the success of these changes. A model for human capital management is essential, both for measuring progress and for outlining specific steps for development (Stahle, 2008; Gogan & Draghici, 2013). Numerous studies confirm the strong correlation between intellectual capital and business performance. Enhancing human capital, as part of intellectual capital, has a substantial positive impact on both financial and organizational performance (Bontis, 1998; Gupta & Roos, 2001; Kalenyuk et al., 2020; Makki & Lodhi, 2018; Guest, 2017), as well as fostering sustainable business growth (Malik et al., 2023; Sitar & Vasic, 2004; Ognjanović, 2016; Nawaz, 2018; Asikhia et al., 2021; Fındık & Husaina, 2020).

Strategies Based on Organizational Culture and Communication. During acquisitions, organizational culture is often preserved in the form of respect for established norms and values. This is because organizational culture is deeply rooted in the national culture of the company and represents the core of the organization. Any changes to the organizational culture are considered the highest level of change. However, in the case of acquisitions by much larger and more successful companies, the adoption of more advanced business processes is inevitable. These processes, often more sophisticated than previous ones, indirectly influence changes in the organizational climate and culture.

Although acquisitions are characterized by the "high speed of implementation," the core values and norms of individuals - along with group norms, values, and beliefs that shape the organizational culture - undergo modernization. Technological advancements, as well as the acquisition of more advanced knowledge, skills, and abilities, lead to changes in the work environment. Business systems such as digital and information systems play a crucial role in this transformation.

Leadership also plays a significant role during the integration of new business processes. Informal authority and subcultures must be considered to increase employee resilience. Effective leadership helps employees adapt to new work systems, business processes, and company values. Employees often hold ingrained habits, beliefs, and opinions, so it is essential that leaders foster change by helping employees acquire new, more sophisticated skills and knowledge.

Change leadership and communication strategies are vital during this phase of the acquisition process. In the case studies, continuous communication from the change leader and their availability for resolving complex situations were crucial. This approach helped create new sets of opinions and attitudes among employees. Additionally, rebranding plays a significant, indirect role in changing organizational culture. By acquiring smaller companies, larger companies can fully position their brand, both in the consumer market and in terms of their employer brand. This implementation of brand values further influences the organization's culture, promoting the desired norms and values.

Communication strategies, which depend on the leader's abilities, personal characteristics, and authentic approach to change management, are key to shifting employees' beliefs, attitudes, and norms. The leader's managerial competencies and cognition also influence the success of these strategies. Furthermore, employees' experiences with their employer - such as fairness in employment relationships, adequate care, protection, and favorable psychological contracts - contribute to increased employee resilience. This, in turn, enhances the effectiveness of managing organizational change.

From the perspective of organizational change management, leaders emphasize the importance of knowledge, abilities, skills, managerial competencies, and leadership skills. Organizational change leaders and human resource managers intersect in their efforts to create a supportive organizational climate and culture, which is essential for effective change management.

According to statistical data, approximately 80% of employees experience cultural tension and competing priorities during periods of change, and struggle to balance them. With top-down communication strategies, only 20% of the workforce fully understands the changes they face. In contrast, "open-source talk" communication strategies help 54% of the workforce understand the changes. The Strategic Human Resources Management (SHRM, 2024) platform also reports that 65% of surveyed managers believe clear and frequent communication is essential for leaders in change management (Fig. 6).

Fig. 6. The organizational change management model and the human resource management model in acquisitions

6 Conclusions and Implication for Further Research

This research highlights the connection between organizational change management and strategic human resource management during transformations, contributing to the further development of both the theoretical and practical concepts of organizational change management. The findings suggest that human resource management strategies, particularly those based on a CSR approach and analyzed through Ulrich's model, have a significant impact on the effectiveness of organizational change management. This influence is particularly evident in the post-acquisition phase and during portfolio project management, aligning with the principles outlined in Kotter's model.

The new model of organizational change management developed in this study refers to the following key conceptual frameworks, which highlight the importance of the board's power and the organization of business systems, through human resource management based on the organizational change management model:

- personal concept of leadership
- balancing strategic and tactical approach
- open-source change management strategies
- flexible organizational structure
- soft skills
- open-source talk

The limitations of this research is very small size of sample, limited literature and narrative approach.

Further research will employ a Grounded Theory methodology, focusing on a sample of change leaders, human resource managers, and decision-makers. The data collected from semi-structured interviews will be analyzed using a system of open coding.

References

Adamović, Ž., Janjić, N., Spasić, D., Ilić, B.: Management of Intellectual Capital in the Company, 9th Conference With International Participation, p. 2023. Knowledge management and informatics, Kopaonik (2023)

Aguinis, H., Kraiger, K.: Benefits of training and development for individuals and teams, organizations and society. Annu. Rev. Psychol. **60**, 451–474 (2009). https://doi.org/10.1146/annurev.psych.60.110707.163505

Ajmal, M.M., Shamim, M., Saleh, M., Malik, A.: Motivation and compensation as predictors of employees' retention: Evidence from public sector oil and gas selling organizations. J. Behav. Sci. **26**(2) (2016)

Al, H.A., Alshurideh, M., Alomari, K., Al Kurdi, B., Alzoubi, H., Hamouche, S., Al-Hawary, S.: The effect of electronic human resources management on organizational health of telecommunications companies in Jordan. Int. J. Data Netw. Sci. **6**, 429–438 (2022). https://doi.org/10.5267/j.ijdns.2021.12.011

Alniacika, E., Almacika, U., Eratb, S., Akcinb, K.: Attracting talented employees to the company: Do we need different employer branding strategies in different cultures? Procedia. Soc. Behav. Sci. **150**, 336–344. 10th International Strategic Management Conference (2014). https://doi.org/10.1016/j.sbspro.2014.09.074

Alniacik, E., Alniacik, U.: Identifying dimensions of attractiveness in employer branding: Effects of age, gender, and current employment status. Procedia. Soc. Behav. Sci. **58**, 1336–1343 (2012). https://doi.org/10.1016/j.sbspro.2012.09.1117

Ali, M.A., Nazimah, H., Hossam, H., Al-A, R., Ibtihal, A.A.: Intellectual Capital and Innovation Performance: Systematic Literature Review. Risks. **9**(9), 170 (2021). https://doi.org/10.3390/risks9090170

Alqudah I. H.A, Carballo-Penela A., Ruzo-Sanmartín E. (2022). High-performance human resource management practices and readiness for change: An integrative model including affective commitment, employees' performance, and the moderating role of hierarchy culture, Eur. Res. Manag. Bus. Econ. 28, 100177. https://doi.org/10.1016/j.iedeen.2021.100177

Anderson, D., Anderson, L.A.: Beyond Change Management. In: How to Achieve Breakthrough Results through Conscious Change Leadership. Google Book, Pfeiffer, San Francisco., e-Book ISBN 978-0-470-64808-7, ISBN 978-0-470-89113-1 (2010)

Antonelli, G., Antonietti, R., Guidetti, G.: Organizational change, skill formation, human capital measurement: Evidence from italian manufacturing firms. J. Econ. Surv. **24**(2), 206–247 (2010). https://doi.org/10.1111/j.1467-6419.2009.00615.x

Asikhia, O.U., Nneji, N.E., Olafenwa, A.T., Owoeye, O.A.: Change management and organisational performance: A review of literature. Int. J. Adv. Eng. Manag. **3**(5), 67–79, ISSN: 2395-5252 (2021) https://doi.org/10.35629/5252-03056779

Barjaktarović S, Cicvarić Kostić S., Kostić-Stanković M. (2021). Communication of the employer brand in job advertisements, UDC: 005.963, Original scientific paper

Bontis, N.: Intellectual capital: An exploratory study that develops measures and models. Manag. Decis. **36**(2), 63–76 (1998) MCB University Press, ISSN 0025-1747

Bozhinovska, T.: Boards' involvement in strategic human resource decisions: Towards an integrative model. J. Hum. Resour. Manag. **12**(1), 70–78 (2019)

Boxall, P., Purcell, J.: Strategic human resource management: Where have we come from and where should we be going? Int. J. Manag. Rev. **2**(2), 183–203 (2000)

Bradford, D.L., Burke, W.W.: Reinventing Organization Development: New Approaches to Change in Organizations, Pfeifer. John Wiley & Sons, Inc., Google Book (2005)

Bradford D.L., Cohen A.R. (2017). Influence without Authority, John Wiley & Sons., https://books.google.rs/books?hl=sr&lr=&id=AkozDwAAQBAJ&oi=fnd&pg=PR9&dq=Brford+D.L.,+Cohen+A.R.+(2017).+Influence+without+Authority&ots=JYJShV6sH4&sig=dUkoSUmaGfoLHVeWWJyypgjlSAU&redir_esc=y#v=onepage&q&f=false

Cummings, T.G., Worley, C.G.: Organization Development & Change, 9th Edition., Library of Congress Control Number: 1234567890, Student Edition ISBN 13: 978-0-324-42138-5, Instructor's Edition ISBN 13: 978-0-324-58054-9, South-Western Cengage Learning, 5191 Natorp Boulevard, Mason, OH 45040. Google Book, USA (2009)

Caputo, A., Molino, M., Cerato, B., Cortese, C.G.: Employer attractiveness: Two instruments to measure employer branding and reputation. SAGE Open. **13**(3), 1–17 (2023). https://doi.org/10.1177/21582440231192188

Changing Point (2024.) https://changing-point.com/ 29/06/2024

Chen, J., Zhu, Z., Xie, H.Y.: Measuring intellectual capital: A new model and empirical study. J. Intellect. Cap. **5**(1), 195–212. Emerald Group Publishing Limited, 1469–1930 (2004). https://doi.org/10.1108/14691930410513003

Chytiri, A.: Human resource manager's role in the digital era. J. Econ. Bus., SPOUDAI Journal. **69**(1–2), 62–72 (2019)

Collins, C.J.: Expanding the resource based view model of strategic human resource management. Int. J. Hum. Resour. Manag. **32**(2), 331–358 (2020). https://doi.org/10.1080/09585192.2019.1711442

Collins, D.: Organizational change: Social Perspectives. Taylor&Francis, Routledge (1998)

Coulter, M.: Strategic Management, Datastatus (2010)

Dahiyat, S.E., Khasawneh, S.M., Bontis, N., Al-Dahiyat, M.: Intellectual capital stocks and flows: Examining the mediating roles of social capital and knowledge transfer. VINE J. Inf. Knowl. Manag. Syst.., Emerald Publishing Limited 2059–5891. (2020). https://doi.org/10.1108/VJIKMS-06-2020-0110

DDI Global Leadership(2021). https://media.ddiworld.com/research/GLF2021-final.pdf

Della Corte V. , Mangia G., Micera R., Zamparelli G. (2008). Strategic Employer Branding: The Brand and Image Management as Attractiveness for Talented Capital., https://ssrn.com/abstract=1960078

Devenport, J., Barrow, S.: Employee Communication During Mergers and Acquisitions. In: The Soft Stuff is the Hard Bit, A Gower Book. Gower Publishing (2009) ISBN: 978-1-3155-7935-1 e-Book

Djordjevic, A., Mihic, M.: Strategic Importance and Sustainable Governance of High-Tech Business Incubators: Evidence from Serbia. In: Mitrovic, N., Mladenovic, G., Mitrovic, A. (eds.) Current Problems in Experimental and Computational Engineering. CNNTech 2021 Lecture Notes in Networks and Systems, vol. 323, pp. 199–221. Springer, Cham (2022). https://doi.org/10.1007/978-3-030-86009-7_11

Dolan, E., Kosasi, S., Sari, S.N.: Implementation of Competence Based Human Resources Management in the Digital Era. Startupreneur Bus. Digit. (SABDA). **1**(2) P-ISSN: 2962-0279, E-ISSN: 2962-0260 (2022). https://doi.org/10.34306/sabda.v1i2.133

Dragicevic, I., Mihic, M.: Women Leadership in Public Sector - Evidence from Serbia. Lex Localis-J. Local Self-Govt. **18**(2), 253–270 (2020). https://doi.org/10.4335/18.2.253-270 (2020)

Dulanović, Ž., Jaško, O.: Organizational Structure and Change, Faculty of the Organizational Sciences, Belgrade (2008)

Eger, L., Micik, M., Gangur, M., Rehor, P.: Employer Branding: Exploring Attractiveness Dimensions ina Multicultural Context. Technol. Econ. Dev. Econ.., ISSN: 2029–4913, eISSN: 2029–4921. **25**(3), 519–541 (2019). https://doi.org/10.3846/tede.2019.9387

Ernst & Young Global Limited. https://www.ey.com/en_gl (2024), Accessed 29 Sep 2024

Eurostat https://ec.europa.eu/eurostat (2024), Accessed 01 May 2024, 26 Jun 2024

Fenech, R.: Human resource management in a digital era through the lens of next generation human resource managers. J. Manag. Inf. Decis. Sci. **25**(S1), 1–10 (2022)

Gartner Inc. https://www.gartner.com/en, (2024) 29/06/2024

Guest, D.E.: Human resource management and organizational psychology. Oxford Res. Encycl. Psychol. (2017). https://doi.org/10.1093/acrefore/9780190236557.013.14

Fernandez-Lores, S., Gavilan, D., Avello, M., Blasco, F.: Affective commitment to the employer brand: Development and validation of a scale. BRQ Bus. Res. Q. **19**, 40–54 (2016). https://doi.org/10.1016/j.brq.2015.06.001

Flotman, A.P., Grobler, A.: The validation of the servant leadership scale. SA J. Ind. Psychol. (2020) ISSN: (Online) 2071–0763

Fountaine, T., McCarthly, B., Saleh, T.: Article Technology Building the AI-Powered Organization, Technology isnt the biggest challange. Culture is. Harv. Bus. Rev. **97**, 62 (2019)

Gartner (2024). https://www.gartner.com/29/06/2024

Gupta, O., Roos, G.: Mergers and acquisitions through an intellectual capital perspective. J. Intellect. Cap. **2**(3), 297–309 (2001) MCB University Press, 1469–1930

Gogan, L.M., Draghici, A.: A model to evaluate the intellectual capital CENTERIS 2013 - Conference on ENTERprise Information Systems / PRojMAN, International Conference on Project MANagement / HCIST 2013 - International Conference on Health and Social Care Information Systems and Technologies. Procedia Technol. **9**, 867–875 (2013)

Hanelta, A., Bohnsack, R., Marzc, D., Maranteb, C.A.: A systematic review of the literature on digital transformation: Insights and implications for strategy and organizational change. J. Manag. Stud. **58**, 5 (2021). https://doi.org/10.1111/joms.12639

Hewagama, G., Boxall, P., Cheung, G., Hutchison, A.: Service recovery through empowerment? HRM, employee performance and job satisfaction in hotels. Int. J. Hosp. Manag. **81**, 73–82 (2019). https://doi.org/10.1016/j.ijhm.2019.03.006

Ivanović, P.: Vademecum of leadership theories Faculty of Economics. Univerzitet u Podgorici (2020)

International Labour Organization https://www.ilo.org/, (2024). Accessed 01 May 2024

Insitute of Margers, Acquisitions and Aliances., https://imaa-institute.org/(2024). Accessed 15 Sep 2024

Jaško, O., Čudanov, M., Jevtić, M., Krivokapić, J.: Organizational Design, Faculty of the Organizational Sciences. Univerity of Belgrade (2013)

Jovanović, M.: Prilog istraživanju uslova za uvođenje agilnih metoda u preduzeća, doktorska disertacija. Univerzitet u Novom Sadu, Fakultet tehničkih nauka (2017)

Jovanović, P.: Investment Management, College of Project Management, Belgrade (2008)

Jovanović, P.: Menadžment - Teorija i praksa, Fakultet organizacionih nauka, Beograd (2005)

Kadima A., Sunardib N., Husaina T. (2020). The modeling firm's value based on financial ratios, intellectual capital and dividend policy, Accounting, 6, (5) 859–870, Growing Science, Canada, doi:https://doi.org/10.5267/j.ac.2020.5.008

Kalenyuk, I., Grishnova, O., Tsymbal, L., Djakona, A., Panchenko, E.: Formation of intellectual corporate capital: Methods and modern trends. Bull. Natl. Acad. Sci. Rep. Kazakhstan., ISSN 1991-3494. **1**(383), 182–191 (2020). https://doi.org/10.32014/2020.2518-1467.23

Kampkötter, P., Mohrenweiser, J., Sliwka, D., Steffes, S., Wolter, S.: IAB Discussion Paper Measuring the use of human resources practices and employee attitudes. In: The Linked Personnel Panel. Institut for Employment Research (2015)

Kaur Bagga, S., Gera, S., Haque, S.: The mediating role of organizational culture: Transformational leadership and change management in virtual teams. Asia Pac. Manag. Rev. **28**, 120–131 (2023). https://doi.org/10.1016/j.apmrv.2022.07.003

Kavanagh, M.H., Ashkanasy, N.M.: The impact of leadership and change management strategy on organizational culture and individual acceptance of change during a merger. Br. J. Manag. **17**(S1), S81–S103 (2006). https://doi.org/10.1111/j.1467-8551.2006.00480.x

Khan, M.I.: The impact of training and motivation on performance of employees. Bus. Rev. **7**(2), 84–95 (2012)

Khana, Z., Soundararajanb, V., Woodc, G., Ahammadd, M.F.: Employee emotional resilience during post-merger integration across national boundaries: Rewards and the mediating role of fairness norms. J. World Bus. **55**, 100888 (2020). https://doi.org/10.1016/j.jwb.2017.06.009

Kinnie, N., Hutchinson, S., Purcell, J., Rayton, B., Swart, J.: Satisfaction with HR practices and commitment to the organisation: Why one size does not fit all. Hum. Resour. Manag. J. **15**(4), 9–29 (2005)

Kotter, J. P.: Leading Change. Harvard Business School Press, Boston (2012)

Kovjanić, M., Vukadinović, P.: Intellectual capital as a strategic performance of the bank with reference to banks. Banking. **50**(1) (2021). https://doi.org/10.5937/bankarstvo2101090K

Kotter, J.P.: Accelerate Building Strategic Agility for a Faster-Moving World, Harvard Business Review Press, Boston, Massachusetts (2014). www.hbr.org

Krstić, B.: Management of intellectual performance of companies. J. Econ. Top., University of Niš. **XLVII**(2), 59–72 (2009)

Lauer, T.: Change Management, Fundamentals and Success Factors. Springer-Verlag GmbH, Berlin Germany., par tog Springer Nature, eBook ISBN 978-3-662-62186-8, ISBN 978-3-662-62187-5 (2021). https://doi.org/10.1007/978-3-662-62187-5

Lee, H.-H., Hong, H.J.S.C.: Human capital and development, lessons and insights from koreas transformation. In: KDI EE Elgar. Korea Development Institute (2018). https://doi.org/10.4337/9781786436979
Leanix (2024). https://www.leanix.net/en/ Accessed 29 Jun 2024
Lengnick-Hall, M.L., Lengnick-Hall, C.A., Andrade, L.S., Drake, B.: Strategic human resource management: The evolution of the field. Hum. Resour. Manag. Rev. **19**, 64–85 (2009)
Li, P., Bastone, A., Ali, M.T., Schiavone, F.: How does artificial intelligence impact human resources performance. evidence from a healthcare institution in the United Arab Emirates. J. Innov. Knowl. **8**, 100340 (2023). ISSN 2444-569X. https://doi.org/10.1016/j.jik.2023.100340
Lievens, F., Slaughter, J.E.: Employer image and employer branding: What we know and what we need to know. Annu. Rev. Organ. Psych. Organ. Behav. **3**, 407–440 (2016)
Makki, M.A.M., Lodhi, S.A.: Impact of corporate governance on intellectual capital efficiency and financial performance. Pakistan J. Commer. Soc. Sci. (PJCSS), Johar Education Society, Pakistan (JESPK) Lahore., ISSN 2309-8619. **8**(2), 305–330 (2014)
Malik, A., Budhwar, P., Mohan, H., Srikanth, N.R.: Employee experience - The missing link for engaging employees: Insights from an MNE's AI-based HR ecosystem. Hum. Resour. Manag. **62**, 97–115 (2023). https://doi.org/10.1002/hrm.22133
Marcyik, G., DeMatteo, D., Festinger, D.: Essentials of Research Design and Methodology. John Wiley & Sons, Inc (2005)
Marjanović, M.: The impact of strategic HRM on change management outcomes. In: Unlocking Hidden Potentials of Organiaztion Through Merging of Humans and Digital, 19th Internaional Symposium, Zlatibor 2024. Univerity of Belgrade, Faculty of Organization Science (2024)
Marjanovic M.: Implementation of Organizational Changes in the Organizational Structure, Bachelors Thesis, Faculty of Organizational Sciences, University of Belgrade (2015)
Mihic, M., Shevchenko, S., Gligorijević, E., Petrovic, D.: Towards Strategic Corporate Social Responsibility Approach in International Projects - Review of South-South Cooperation: A Case Study of Chinese Projects in Angola. Sustainability. **11**(10), 2784 (2019). https://doi.org/10.3390/su11102784
Monteiro, B., et al.: Employer branding applied to SMEs: A pioneering model proposal for attracting and retaining talent. Information. **11**, 574 (2020). https://doi.org/10.3390/info11120574
Mubarik, M.S., Chandran, V.G.R., Devadason, E.S.: Measuring human capital in small and medium manufacturing enterprises: What matters? Soc. Indic. Res. (2017). https://doi.org/10.1007/s11205-017-1601-9
Nawaz, T.: Intellectual capital profiles and financial performance of Islamic banks in the UK. Int. J. Learn. Intellect. Cap. Inderscience (2018). https://doi.org/10.1504/IJLIC.2019.096934
Nielsen, K., Randall, R., Christensen, K.B.: Does training managers enhance the effects of implementing team-working? A longitudinal, mixed methods field study. Hum. Relat. **63**(11), 1719–1741 (2010). https://doi.org/10.1177/0018726710365004
Nurlaila, P.: Performance model: Satisfaction, commitment and reward based. KINERJA: J. Manaj. Organ. Ind. **1**(1), 19–26 (2022)
Nurzaman, A.M.E., Affandi, A., Udobong, A., Hernawan, S.: Implementation of human resource management in the adaptation period for new habits. Int. J. Educ. Admin. Manag. Leadersh. **1**(1), 19–26 (2020) ISSN (Print): 2580–1309
Obradović, V., Petrović, D., Mihić, M.: Identification of changes: The first step. In: Operational management in the function of sustainable economic growth and development of Serbia 2011–2020. Faculty of Organizational Sciences in Belgrade (2011)
Ocak, M., Fındık, D.: The impact of intangible assets and sub-components of intangible assets on sustainable growth and firm value: Evidence from Turkish listed firms. Sustainability. **2019**(11), 5359 (2019). https://doi.org/10.3390/su11195359

Petrović D, Mihić M, Stošić B. (2008). Strategic IT portfolio management for development of innovative competences, in Handbook on Strategic Information Technology and Portfolio management (eds). Tan A., Theodorou P., IGI Publishing, Information Science Reference, Hershey, USA, ISBN 978-1-59904-687-7, pgs. 150–169.

Tornjaski, V.Ž.: Predicting the success of organizational changes with the quantitative Beckard Harris model. In: Doctoral Dissertation. Faculty of Organizational Sciences, University of Belgrade (2019)

Sabie, O.M., Brişcariu, R.M., Pîrvu, C., Burcea, S.G., Gâtan, M.L.: The relationship berween emotional intelligence and human resources employee performance: A case study for romanian companies. Manag. Res. Pract. **1 2**(3) (2020) ISSN 2067-2462

Salas, E., Diaz, G.D., Klein, C., Burke, C.S., Goodwin, S., Halpin, G.: Does team training improve team performance? A meta-analysis. Hum. Factors. **50**(6), 903–933 (2008). https://doi.org/10.1518/001872008X375009

Salvatore, D.: International Economics. Faculty of Economics, University of Belgrade (2014)

Samoliuk, N., Bilan, Y., Mishchuk, H., Mishchuk, V.: Employer brand: Key values influencing intention to join the company. Manag. Mark. Chall. Knowl. Soc. **17**(1), 61–72 (2022). https://doi.org/10.2478/mmcks-2022-0004

Sancak, E.: Change management in sustainability transformation: A model for business organizations. J. Environ. Manag. **330**, 117165 (2023). https://doi.org/10.1016/j.jenvman.2022.117165

Schellinger, J., Tokarski, K.O., Kissling-Näf, H.I.: Digitale Transformation und Unternehmensführung Trends und Perspektiven für die Praxis., ISBN 978-3-658-26959-3, ISBN 978-3-658-26960-9 (eBook). Springer (2020). https://doi.org/10.1007/978-3-658-26960-9

Scott-Jackson, W., Mayo, A.: HR with Purpose: Future Models of HR. Henly Business Scholl (2018)

Shrivastava, N.A.: Adopting HR analytics: Factors influencing Human Resource Professionals. Int. J. Res. Anal. Rev. (IJRAR). **11**(1) (2024) www.ijrar.org (E-ISSN 2348–1269, P-ISSN 2349–5138)

SHRM. https://www.shrm.org/ (2024). Accessed 01 May 2024, 29 Jun 2024

Simić, M., Ilić, M.: Informal communication and conflicts. Example Serbia Kargo ad, UDK. **656**(2), 658+658.8 (2019)

XIX International Simposium Symorg24. https://symorg.fon.bg.ac.rs/ (2024), Accessed 12–15 Jun 2024

Sitar A. S., Vasic V. (2004). Measuring Intellectual Capital: Lessons Learned from a Practical Implementation

Slavković, M., Ognjanović, J.: Impact of human capital on business performance of hotel enterprises in Serbia. TEME. **XLII**(4), 1339–1355 (2018). https://doi.org/10.22190/TEME1804339S

Snape, E., Redman, T.: An evaluation of a three-component model of occupational commitment: Dimensionality and consequences among united kingdom human resource management specialists. J. Appl. Psychol. **88**(1), 152–159 (2003). https://doi.org/10.1037/0021-9010.88.1.152

Stahle, P.: National intellectual capital as an economic driver. In: Perspectives on Identification and Measurement (2008)

Stanivuk, M.: Intellectual capital in the sigital economy. Economics. **3**(1) ISSN 2303-5005 (2015). https://doi.org/10.1515/eoik-2015-0003

Stouten, J., Rousseau, D.M., Cremer de, D.: Successful organiazctional change: Integrating the management practice and scholarly literatures. Acad. Manag. Ann. **12**(2), 752–788 (2018). https://doi.org/10.5465/annals.2016.0095

Shuler, R.S.: Strategic human resources management: Linking the people with the strategic needs of the business. Organ. Dyn. **21**(1), 18–32 (1992)

Supply Chain Digital. https://supplychaindigital.com/ (2024), Accessed 15 Sep 2024

S&P Global. https://www.spglobal.com/ (2024), Accessed 15 Sep 2024

Ulrich, D., Grochowski, J.: Building a world class hr department. Strateg. HR Rev. (2018). https://doi.org/10.1108/SHR-05-2018-0046

Waddell, J.T.: Servant leadership. Servant Leadership Research Roundtable. (2006) https://regentparents.regent.edu/acad/global/publications/sl_proceedings/2006/waddell.pdf

World Bank. https://www.worldbank.org/en/home (2024), Accessed 01 May 2024

World Heth Organization. https://www.who.int/ (2024), Accessed 01 May 2024

Zavod za statistiku Republike Srbije. https://data.gov.rs/sr/ (2024), Accessed 01 May 2024, 26 Jun 2024

Zhang, X., Lin, Z., Liu, Y., Chen, X., Liu, D.M.: How do human resource management practices aect employee well-being? a moderated mediation model. Empl. Relat. **42**(4), 903–919 (2020)

Academic Support and Youth Project Engagement Across Serbia's Regions: "If you build it, they will come"

Dragan Bjelica(✉), Dejan Petrović, and Petar Stanimirović

Faculty of Organizational Sciences, University of Belgrade, Belgrade, Serbia
dragan.bjelica@fon.bg.ac.rs

Abstract. The research presented in this chapter examines the impact of higher education **institutional support** on the participation of **youth**, aged 18-24, in **extracurricular activities** across the Republic of Serbia and its neighbors (Montenegro and BiH), with particular attention to regional disparities. The research encompasses a sample of 1,638 university **students** from mentioned regions. Differences between groups were analyzed using Kruskal-Wallis tests. The findings underscore the significant role of extracurricular activities, such as involvement in **student projects** and participation in **student organizations**, across all regions. Notably, students from less developed areas, specifically Eastern and Southern Serbia, exhibit a higher propensity for engaging in these activities. Furthermore, the study reveals that extracurricular engagement is not influenced by academic performance. These results provide valuable insights for formulating policy recommendations and developing programs within **higher education institutions**.

Keywords: Higher Education Institution · Project Engagement · Youth · Youth Engagement · Support Activities · Institutional Support · Extracurricular Activities · Students · Student Projects · Student Organization · Life Satisfaction

1 Introduction

Higher education institutions hold a pivotal role in society, as they are expected to provide various programs that foster continuous development of young people. Proactive higher education system must not only respond to current educational needs but also anticipate future labor market trends and the evolving expectations of youth in order to create modern and effective study curricula and extracurricular programs. These programs should include student projects, participation in student organizations, debates, case study solutions, and various student competitions. Throughout history, the prosperity of prominent civilizations, such as Ancient Greece, has been closely tied to their investment in the education and intellectual development of their citizens. Similarly, in contemporary educational systems whose key to success lies in building robust support programs and systematic approaches that empower students to reach their full potential. When opportunities and resources are properly aligned, students naturally gravitate towards them, eager to grow and succeed.

V. Štavljanin et al. (Eds.): SymOrg 2024, LNNS 1681, pp. 166–181, 2025.
https://doi.org/10.1007/978-3-032-08090-5_12

However, Research by Rubio et al. [48] identifies financial challenges, resource inequalities and insufficient knowledge as significant barriers that many students face during their schooling at modern higher education institutions. In order to overcome these obstacles, it is important that higher education institutions develop support programs, especially those that assist students transitioning from secondary school and also those that are fostering their project participation. Tailored onboarding programs, as Penn-Edwards & Donnison [40] point out, are crucial in ensuring a smooth transition, especially during the first week of university, when students are introduced to essential resources and guidance. These programs can articulate students' expectations and contributes to lowering dropout rates [27]. Additionally, involving youth in participatory research and evaluation yield significant insights but also it empowers them, resulting in both individual and societal benefits [44]. Furthermore, extracurricular entrepreneurial activities can foster entrepreneurial self-efficacy, shaping students' interests and aspirations [45].

Accordingly, engaging students in extracurricular activities and projects within higher education institutions plays a pivotal role in their overall development and academic success [28]. Youth participation in these activities is a topic that has garnered significant research interest. Study conducted by Brown & Evans [8] demonstrates that participation in extracurricular activities positively impacts academic performance, while Bjelica et al. [4] further emphasize the link between project involvement and **life satisfaction** among students. In particular, community-based projects promote civic responsibility and help students form supportive networks outside their immediate social circles, which act as protective factors against negative behaviors [12]. Additionally, students who take on leadership roles in extracurricular settings develop essential leadership skills and self-efficacy, enhancing both personal and professional prospects [20]. Furthermore, Gajović et al. [19] highlight the importance of environmental engagement in fostering youth awareness of sustainability and global challenges.

Collectively, mentioned facts underline the multifaceted factors influencing positive outcomes for young people, highlighting the need for further research into specific elements that encourage youth participation in projects in Serbia. This topic holds significant relevance for developing countries, such as those in the Western Balkans, where numerous public policies focus on youth. A notable example is the Youth Strategy of the Republic of Serbia until 2030, which prioritizes youth engagement and participation as one of main development goals. However, there is a scarcity of research addressing youth engagement, particularly in the context of project-based initiatives within developing countries. To address identified gap this chapter, therefore, investigates the involvement of young individuals in higher education institutions and analyzes their perceptions of different support activities within this context. The aim is to identify effective strategies for formulating policies and programs that enhance student engagement and life satisfaction.

2 Literature Review

The life quality and the life satisfaction of youngsters are crucial indicators of societal progress and national prosperity. Life satisfaction, as defined by subjective perceptions of well-being and fulfillment, provides insight into how individuals assess their overall life circumstances [14]. When it comes to the Serbia, as in many other states, the education system plays a vital component influencing young people's lives and is recognized as a key factor affecting their overall quality of life [43]. Accordingly, in the following text, the key findings of current scientific literature dealing with the topic of engaging young people in projects at higher education institutions will be presented in order to gain a deeper understanding of the issue and the whole scientific area of the conducted research.

2.1 Importance of Youth Extracurricular Engagement

In recent years, research on youth engagement has underscored the significant role that both academic and extracurricular activities play in enhancing overall life quality and satisfaction among young people [15, 34, 51]. Extracurricular involvement not only supplements traditional education and skill development but also contributes to the formation of socially adaptable and emotionally resilient individuals. This type of engagement provides young people with opportunities for self-definition and goal achievement, ultimately improving their life satisfaction [15].

Personal development through extracurricular activities has been shown to be crucial for fostering positive behavioral and academic outcomes as well as for strengthening social bonds, daily responsibilities, and personal achievements [3]. Heath et al. [22] and Schwartz et al. [50] demonstrate that students involved in extracurricular activities achieve higher academic success, reflected in better test scores and higher grades, alongside enhanced self-confidence.

Findings from Martinez et al. [28] suggest that participation in activities like sports, clubs, and the arts has a favorable impact on students' perceptions of their school environments, although the effects may vary based on the specific combination of activities. Similar findings were emphasized in work of Fredricks and Eccles [18]. Notably, students who engage in sports tend to exhibit stronger school engagement and experience greater developmental benefits, as demonstrated by Forneris et al. [17], who also highlight the importance of support from parents and adult mentors in encouraging diverse extracurricular involvement.

Furthermore, research by Bundick [9] illustrates a clear link between participation in extracurricular activities and positive developmental outcomes, particularly through involvement in leadership and volunteer roles. In addition, interpersonal relationships built trough connections with peers, teachers, and parents have consistently been associated with improved behavioral outcomes, school attachment, and overall well-being, further solidifying the importance of fostering social relationships in the educational setting [39]. To this end, Bjelica and Jovanovic [3] advocate for higher education institutions to introduce programs that promote youth well-being by encouraging extracurricular participation.

These findings collectively suggest that both extracurricular and academic engagements play a fundamental role in the holistic development of young individuals, offering them valuable life experiences, social connections, and an enhanced sense of responsibility.

2.2 Extracurricular Activities and Youth Participation in Projects

Extracurricular activities for youth are frequently motivated by opportunities for volunteering, which provide valuable experience and knowledge pertinent to project work [46]. Such engagements also reflect a broader desire among young people to address community issues. Peréa et al. [41] note that involvement in social projects fosters the development of research and communication skills, enhancing their connection to civic initiatives. This kind of participation can help establish a sustainable system of project involvement that aligns with both personal interests and community needs through effective networking [11].

In the Republic of Serbia, however, youth participation in extracurricular activities is influenced by several factors including population density, GDP, employment rates, and demographic characteristics [34]. Pavlović et al. [37] highlight the need to address regional disparities and varying behavioral patterns among youth in different areas of Serbia. The availability of extracurricular opportunities often depends on local conditions and the ability of educational institutions to offer such activities, emphasizing the importance of equitable access across diverse regions.

There are numerous types of projects in which youth can participate, and engagement in one form of activity often complements involvement in others. Schelven et al. [49] highlight how participation in panels and special events helps youths develop organizational skills and competencies crucial for short-term tasks. Similarly, Bianco et al. [2] emphasize the value of volunteerism and involvement in educational and student-initiated **projects**, which provides practical experience within institutional settings.

Research projects are also noted for their role in skill development. Smith et al. [52] underscore the importance of projects that enhance data analysis and presentation skills, which are vital for academic and professional success. In addition, strategic collaborations between educational institutions and businesses, as discussed by Bjelica et al. [5] and Medina et al. [29], offer youths practical experience and valuable networking opportunities. Wagner and Compton [57] explore how startup initiatives during schooling can foster entrepreneurial skills among young people. Mentoring programs and career guidance, examined by Ilieva and Vitanova [24] and Meltzer et al. [30], are also essential in facilitating the transition to the workforce by addressing both organizational and individual needs.

Projects aimed at supporting individuals with special needs, as discussed by Kramer et al. [26], help enhance empathy and support among participants. Additionally, Novytskyi [35] highlights the benefits of sports projects in promoting a healthy lifestyle, while Ozer and Douglas [36] and Suraswadi [54] focus on community-oriented projects that address local issues. Collectively, these diverse forms of engagement provide youths with invaluable experiences and skills, preparing them for various future roles and contributing to their comprehensive development.

2.3 Higher Education Institutions Support for Youth Projects Engagement

To facilitate student participation in extracurricular activities, higher education institutions can implement programs that offer mentorship support, create incentive and effective evaluation systems and provide career counseling to address any potential adverse effects on academic performance [1]. Partnerships during this process with various stakeholders such as student organizations can further streamline the identification and coordination of pertinent activities, thereby promoting leadership development [25].

One example of good practice was presented in work of Paek et al. [38]. The authors propose improving the academic performance of undergraduate Computer Science students by implementing an "Innovation, Competency, and Experience score (ICE)", enabled through a web platform named "RadGrad" [38]. This approach promotes and incentivizes extracurricular activities engagement, underscoring the importance of a comprehensive evaluation that includes experiences relevant to students' careers in addition to traditional course assignments.

Enhancing the experience of students and advancing their overall academic success can be significantly achieved through the development of transferable competencies facilitated by extracurricular activities and project-based engagement [23]. Also, empirical evidence of same authors indicates a positive correlation between participation in extracurricular activities and academic performance, with the relationship being mediated by reduced academic stress and enhanced learning efficiency [56]. Such activities offer crucial opportunities for the advancement of individual talents, exploration of students' interests, and the acquisition of distinctive skills, which collectively foster both personal development and academic achievement. Nonetheless, the effective administration and structuring of extracurricular programs by educational institutions are pivotal to their successful execution [16].

Based on the literature review conducted for this research, a list of effective activities aimed at fostering youth engagement in projects at higher education institutions has been identified. According to research conducted by Bjelica et al. [5], these activities include [5]:

- Allocating substantial time for academic study and scholarly activities,
- Offering targeted support mechanisms to facilitate student academic achievement,
- Utilizing academic support services such as reading rooms, libraries, and laboratories,
- Encouraging interactions among students from varied socio-economic backgrounds, racial identities, and religious affiliations,
- Providing resources designed to enhance students' overall well-being, including recreational facilities, healthcare services, and counseling support,
- Supporting the management of extracurricular activities alongside work and family responsibilities,
- Engaging in forums and discussions focused on contemporary social, economic, and political issues.

Fostering youth engagement in extracurricular projects within higher education institutions through the implementation of the outlined support activities is essential for both personal development and academic achievement. Moreover, these activities provide students with transferable skills critical for future career success. Through strategic planning and collaboration, institutions can ensure that extracurricular involvement becomes a vital mechanism for shaping well-rounded, skilled, and motivated individuals [31].

The agile methodology is the most suitable approach for managing youth projects, as it emphasizes flexibility and iterative progress [42]. Engaging young people actively in projects is a crucial factor in achieving overall project success, as their involvement can significantly enhance project outcomes and effectiveness [53, 55].

3 Methodology

This research is funded by the Science Fund of the Republic of Serbia under the project titled "Engagement in Academic Achievements and Extracurricular Activities as Predictors of Life Satisfaction among High-School and University Students - SHINE" which was conducted during 2022 and 2023 through the collaboration of three institutions: University of Belgrade's Faculty of Organizational Sciences, Faculty of Law and the Institute of Economic Sciences in Belgrade [5].

Part of the research presented in this chapter sought to investigate the impact of institutional support endeavors on youth extracurricular engagement across the Republic of Serbia and its neighbors (Montenegro and BiH), with a particular emphasis on their participation in student projects and its correlation with their residential location. The participants were students from various universities across Serbia, categorized into six groups of their origin: Belgrade, Vojvodina, Kosovo and Metohija, Southern and Eastern Serbia, Šumadija and Western Serbia and Foreign (Montenegro, BiH) [5].

An online questionnaire, developed using Google Forms for research purposes, was employed as the primary research instrument. The questionnaire consisted of several segments, including section for collecting demographic data (e.g. age, gender, grade point average - GPA, place of residence), and sections exploring attitudes regarding the significance of youth engagement in projects at higher education institutions, as well as perceptions of available institutional support activities for such engagement. While demographic questions were presented as multiple-choice, the attitude-related questions were structured on a four-point Likert scale: 1 - "Undecided"; 2 - "I don't plan to"; 3 - "I plan to"; 4 - "I have already engaged or am currently engaged." Additionally, a series of open-ended questions provided respondents with the opportunity to articulate their perspectives in greater detail [5].

The questionnaire was administered to students during the teaching process, resulting in a total of 1,638 completed responses, which constituted the research sample. Analysis of collected data was performed using SPSS 29.0 software, with cross-tabulations used to examine demographic patterns in relation to levels of institutional support. Additionally, the Kruskal-Wallis tests were employed to identify significant differences between the groups defined in the study. The findings, which provide valuable insights, are thoroughly detailed in the section of this chapter dedicated to research results, while the demographic characteristics of the sample are presented in the following Table 1.

4 Research Results

This study forms part of a broader research initiative conducted under mentioned SHINE project. In both this study and the larger research effort, various types of extracurricular activities were examined to assess youth project engagement. These activities were analyzed based on respondents place of residence, with a focus on the following factors

Table 1. Research sample descriptive statistics

Place of residence		Gender			Student organization member			GPA
		Male	Female	Total	Yes	No	Total	
Belgrade Region	Count	291	622	913	240	673	913	8.15
	% within Residence	31.9%	68.1%	100.0%	26.3%	73.7%	100.0%	
	% within Gender	59.5%	54.1%	55.7%	57.6%	55.1%	55.7%	
	% of Total	17.8%	38.0%	55.7%	14.7%	41.1%	55.7%	
Vojvodina Region	Count	27	73	100	32	68	100	8.32
	% within Residence	27.0%	73.0%	100.0%	32.0%	68.0%	100.0%	
	% within Gender	5.5%	6.4%	6.1%	7.7%	5.6%	6.1%	
	% of Total	1.6%	4.5%	6.1%	2.0%	4.2%	6.1%	
Šumadija and Western Serbia	Count	114	311	425	93	332	425	8.18
	% within Residence	26.8%	73.2%	100.0%	21.9%	78.1%	100.0%	
	% within Gender	23.3%	27.1%	25.9%	22.3%	27.2%	25.9%	
	% of Total	7.0%	19.0%	25.9%	5.7%	20.3%	25.9%	
Southern and Eastern Serbia	Count	49	124	173	46	127	173	8.22
	% within Residence	28.3%	71.7%	100.0%	26.6%	73.4%	100.0%	
	% within Gender	10.0%	10.8%	10.6%	11.0%	10.4%	10.6%	
	% of Total	3.0%	7.6%	10.6%	2.8%	7.8%	10.6%	
Kosovo and Metohija	Count	5	10	15	4	11	15	8.16
	% within Residence	33.3%	66.7%	100.0%	26.7%	73.3%	100.0%	
	% within Gender	1.0%	0.9%	0.9%	1.0%	0.9%	0.9%	
	% of Total	0.3%	0.6%	0.9%	0.2%	0.7%	0.9%	

(*continued*)

[5]: 1. Participation in internal business practice programs within companies (**internal**); 2. Participation in student organizations or groups (**stud_org**); 3. Participation in joint learning programs or formal programs where groups of students attend one or more classes together (**joint_learn**); 4. Participation in school or student projects (**stud_proj**);

Table 1. (*continued*)

Place of residence		Gender			Student organization member			GPA
		Male	Female	Total	Yes	No	Total	
Other (Montenegro, BiH)	Count	3	9	12	2	10	12	8.12
	% within Residence	25.0%	75.0%	100.0%	16.7%	83.3%	100.0%	
	% within Gender	0.6%	0.8%	0.7%	0.5%	0.8%	0.7%	
	% of Total	0.2%	0.5%	0.7%	0.1%	0.6%	0.7%	
Total	Count	489	1149	1638	417	1221	1638	8.18
	% within Residence	29.9%	70.1%	100.0%	25.5%	74.5%	100.0%	
	% within Gender	100.0%	100.0%	100.0%	100.0%	100.0%	100.0%	
	% of Total	29.9%	70.1%	100.0%	25.5%	74.5%	100.0%	

5. Participation in solving business case studies **(case_stud)**; 6. Participation in school or student academic competitions **(stud_comp)**; 7. Participation in debates and public discussions **(stud_deb)**; 8. Participation in student exchange programs **(stud_exch)**; 9. Participation in specialized seminars, professional gatherings, and events **(stud_sem)**; 10. Collaborative work with teaching staff on competition preparation **(wp_comp)**; and 11. Collaborative work with teaching staff on research or commercial projects **(wp_proj).**

Overall, the highest levels of youth engagement were observed in student organizations **(stud_org)** and school or student projects **(stud_proj)**. Additionally, the research highlights a higher level of engagement in extracurricular activities among youth from Kosovo and Metohija, as well as Southern and Eastern Serbia what is illustrated in Fig. 1. According to data from the Republic Institute of Statistics, these regions are considered less developed which means that this finding may hold significance for decision-makers in local self-government units within these areas, as it suggests that young people from these regions possess considerable potential, which is often realized in more developed regions.

Given the significance of student engagement in projects, recognized as one of the most highly valued extracurricular activities, the analysis centered on youth participation across different institutional measures of support. Although project management is predominantly introduced at the university level, initiatives implemented in elementary schools have also played a role in promoting more structured thinking among younger students [13].

In the study, students were asked to assess the degree to which their institution emphasizes the support measures, in a manner similar to the broader research conducted under the SHINE project [5]. These measures are [5]: a. Participation in forums and gatherings

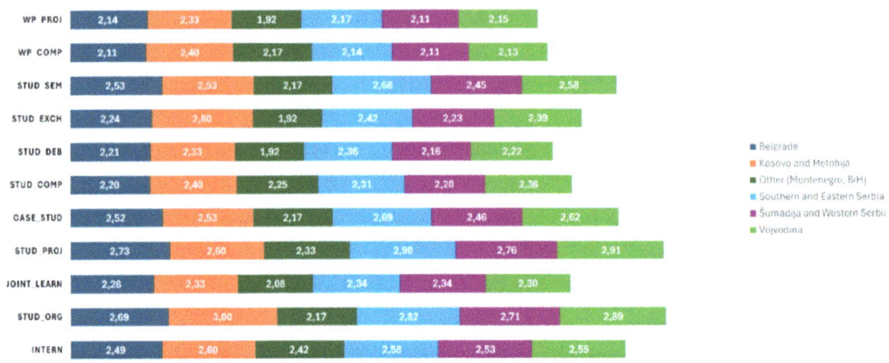

Fig. 1. Students' engagement in university extracurricular activities attitudes based on their place of residence

dedicated to contemporary social, economic, and political issues - **inst_events**; b. Assistance in managing extracurricular commitments and obligations, including work and familial responsibilities - **inst_extracurr**; c. Provision of resources aimed at enhancing the overall well-being of students, encompassing recreational activities, healthcare provisions, and counseling services - **inst_welf**; d. Providing support to students to achieve academic success - **inst_accsup**; e. Use of learning support services (reading room, library, laboratory) - **inst_serv**; f. Promoting interactions among students from diverse socio-economic backgrounds, racial identities, religious affiliations, etc. - **inst_netw**; g. Allocating a significant amount of time to study and academic work - **inst_learn.**

The findings were further validated through Kruskal-Wallis statistical tests. In general, the mean values show that institutional support activities were rated at different levels by the respondents. Activities related to inst_welf (3.04 ± 1.274) and inst_extracurr (2.97 ± 1.297) received the lowest scores among all categories, suggesting that students perceive these aspects as the least supported or important. In contrast, inst_serv was rated the highest (3.92 ± 1.132), indicating that institutional services such as reading rooms, libraries, and laboratories are recognized as the most valuable and effective by students.

Using the Kruskal-Wallis test (Table 2), significant differences between the groups were observed for most institutional support parameters. Specifically, significant differences were found for the following: inst_learn ($p = 0.018$), inst_accsup ($p = 0.008$), inst_serv ($p = 0.010$), inst_extracurr ($p = 0.027$). This indicates that the evaluation of these activities varies significantly among different respondents from different regions. Conversely, no statistically significant differences were found for inst_netw ($p = 0.205$) and inst_welf ($p = 0.177$), suggesting that these support measures are equally valued regardless of the respondents' place of residence.

The same statistical test was applied to analyze differences in institutional support parameters across GPA groups. Respondents were categorized into three groups based on their GPA, these gropes are: GPA under 7.5, GPA between 7.5 and 8.5, and GPA higher than 8.5. The analysis yielded no significant differences in students' engagement in extracurricular activities based on their GPA. This suggests that the level of student

Table 2. Analysis of educational institutional support in relation to student place of residence

Place of residence		inst_learn	inst_accsup	inst_serv	inst_netw	inst_welf	inst_extracurr	inst_events
Belgrade Region	Mean	3.59	3.53	3.91	3.23	3.02	2.92	3.40
	N	913	913	913	913	913	913	913
	Std. Dev.	1.066	1.126	1.096	1.310	1.260	1.307	1.238
Vojvodina Region	Mean	3.53	3.44	3.86	3.23	3.02	2.88	3.36
	N	100	100	100	100	100	100	100
	Std. Dev.	1.167	1.192	1.206	1.347	1.326	1.328	1.251
Šumadija and Western Serbia	Mean	3.66	3.57	3.85	3.29	3.09	3.00	3.38
	N	425	425	425	425	425	425	425
	Std. Dev.	1.032	1.129	1.220	1.306	1.266	1.259	1.237
Southern and Eastern Serbia	Mean	3.83	3.84	4.17	3.51	3.06	3.16	3.68
	N	173	173	173	173	173	173	173
	Std. Dev.	.971	1.010	1.020	1.279	1.317	1.278	1.224
Kosovo and Metohija	Mean	4.07	3.87	4.40	3.53	3.47	3.73	3.60
	N	15	15	15	15	15	15	15
	Std. Dev.	.799	.990	.910	.834	1.246	1.335	1.242
Other (Montenegro, BiH)	Mean	3.17	3.00	3.50	3.08	2.17	2.58	3.00
	N	12	12	12	12	12	12	12
	Std. Dev.	.937	1.279	1.243	1.505	1.337	1.443	1.414
Total	Mean	3.63	3.57	3.92	3.28	3.04	2.97	3.42
	N	1638	1638	1638	1638	1638	1638	1638
	Std. Dev.	1.054	1.123	1.132	1.307	1.274	1.297	1.240
Kruskal-Wallis H test		13.679	15.612	14.979	7.218	7.642	12.658	10.660
df		5	5	5	5	5	5	5
Asymp. Sig.		.018*	.008**	.010**	.205	.177	.027*	.050*

* significance at the level 0.05; ** significance at the level 0.01

involvement in extracurriculars is not influenced by academic performance, as measured by their grade point average.

5 Discussion

This research identifies key factors influencing youth engagement in projects, with particular emphasis on the role of educational institutions in providing support. The results indicate that adequate institutional support, access to university resources, infrastructure, and mentoring programs significantly enhance youth participation and motivation to engage in projects. Similar findings, emphasizing the importance of mentoring for student engagement, are reported by Ilieva and Vitanova [24] and Meltzer et al. [30]. Additionally, Hanushek & Woessmann [21]. These papers highlight a positive correlation between school resources and academic success. On the other hand, the research conducted by Burtless, G [10]. showed that there is no statistically significant relationship between academic success and project engagement.

Accordingly, higher-education institutions can enhance youth project engagement by offering personalized mentorship and specialized resources aligned with youth interests for all students regardless of their university grades. Development of programs that bridge theoretical knowledge with practical experience may better prepare young people for project-related challenges. This research proposes that educational institutions can serve as catalysts for active participation by incorporating relevant projects into curricula, thereby equipping young people with essential professional skills. Implementing these recommendations could further strengthen the effectiveness of educational programs, improve youth involvement in social initiatives and their quality of life [15, 32, 51].

When supporting projects to engage young people, particular attention should be given to students from less developed regions. The research indicates that these students demonstrate notably higher involvement in such initiatives compared to their peers from more developed areas. Based on this finding, local governments in these regions should prioritize support for these students, creating incentives for them to return to their hometowns and fulfill their potential locally. Such efforts could contribute to accelerating the development and progress of these regions. Similar findings have been confirmed in countries in the region such as Croatia by Botrić V [7]., while a modern approach to youth engagement that supports this trend has been elaborated in Youniss, J 's [58] work.

Tu sum up, the findings of the research fully substantiate the initial hypothesis that greater student project engagement enhances youth satisfaction and overall quality of life. These insights hold practical implications for educational institutions and youth-focused organizations, particularly those aiming to foster greater project involvement among young people. This will significantly improve the prerequisites for the successful development and implementation of the knowledge-based approach to project management that the authors Todorovic, M. et al. [54] and Rošulj, D. et al. [47] wrote about, and which, as one of the key critical success factors, significantly increases the chances of project success. Furthermore, Miković et al. [33] highlight that fostering youth social capital can play a pivotal role in advancing a knowledge-based management system within the non-profit sector. Strengthening young people's capacities in project management - particularly through active participation in projects led by student and other educational non-profit organizations - can significantly enhance the competencies of future project managers. This approach contributes to the cultivation of leaders who will drive the success of projects with transformative impacts on the economy and society, thereby supporting the development of a knowledge-based social system which is also discussed in a similar way by the authors Borozan T. et al. [6].

However, several limitations of this study should be noted when interpreting the findings. First, the study's focus on a specific geographic context may restrict the generalizability of results to regions with different educational and social characteristics. Additionally, the methodology relies on self-reported data, which may introduce bias due to participants' subjective perceptions of support and personal engagement. In light of these limitations, future research should include broader demographic and geographic samples to gain deeper insights into variations among youth groups. Researchers might also consider employing methodologies that involve direct observation of project engagement and interviews with mentors and educators to obtain more objective data on the effects of institutional support. Additionally, future studies could investigate specific types of institutional programs that most effectively foster youth engagement, enabling the development of more targeted recommendations for educational institutions.

6 Conclusion

Higher education institutions in the Republic of Serbia are pivotal in influencing the career trajectories of young individuals. This research, supported by the Science Fund of the Republic of Serbia through the Social Sciences and Humanities Program, Ideas, aims to investigate the disparities and requirements of youth engaged in extracurricular activities.

The study reveals that participation in student projects and related activities is highly esteemed among young people. Interestingly, students from less developed regions demonstrate a stronger inclination towards project involvement. Additionally, the perception of institutional support varies significantly depending on the student's region of residence. However, the research indicates no correlation between extracurricular engagement and Grade Point Average (GPA). Instead, students actively involved in projects and other extracurricular endeavors report a heightened perception of institutional support.

These findings offer significant insights for policymakers and educational leaders, providing a foundation for refining support structures to better address the diverse needs of students across regions. However, the research has limitations that may affect the generalizability of the results, including its geographic specificity. Additionally, the reliance on self-reported data introduces potential subjectivity, suggesting the value of integrating more objective data collection methods. Future research should aim to include geographically varied samples and employ diverse data collection techniques to validate these findings and enrich understanding of institutional support dynamics.

Acknowledgements. This research was supported by the Science Fund of the Republic of Serbia, under project number 7754305, titled "Engagement in Academic Achievements and Extracurricular Activities as Predictors of Life Satisfaction among High School and University Students - SHINE."

References

1. Almasry, M., et al.: Perceptions of preclinical medical students towards extracurricular activities. Int. J. Med. Educ. **8**, 285 (2017)
2. Bianco, V., Cater, M., Fox, J.: 4-h youth volunteer retention in science projects. J. Youth Dev. **16**(4), 134–148 (2021). https://doi.org/10.5195/jyd.2021.1004
3. Bjelica, D.L., Jovanović, U.D.: It's up to you: The influence of sports participation, academic performances and demo-behavioral characteristics on university students' life satisfaction. Appl. Res. Qual. Life. **11**(1), 163–179 (2016). https://doi.org/10.1007/s11482-014-9360-0
4. Bjelica, D., Pavlović, D., Petrović, L.: Beyond the GPA: Factors influencing life satisfaction of engineering undergraduates. Int. J. Eng. Educ. **40**(3) (2024)
5. Bjelica, D., Petrović, D., Stanimirović, P.: Youth Project Engagement: The Role Of Educational Institutional Support. In: XIX International Symposium SymOrg 2024 - Unlocking The Hidden Potentials Of Organization Through Merging Of Humans And Digitals, pp. 813–817. Faculty of Organizational Sciences, University of Belgrade, Zlatibor, Serbia, 12-15 June 2024 (2024). isbn:978-86-7680-464-1
6. Borozan, T., Maričić, M., Bjelica, D.: Impact of Business Simulation Games on Skills of Engineering Undergraduates. Int. J. Eng. Educ. **40**(5), 1090–1100 (2024)
7. Botrić, V.: Urban–rural differences in youth engagement in Croatia. Reg. Sci. Policy Pract. **15**(4), 909–927 (2023)
8. Brown, R., Evans, W.: Extracurricular activity and ethnicity. Urban Educ. **37**(1), 41–58 (2002). https://doi.org/10.1177/0042085902371004
9. Bundick, M.J.: Extracurricular activities, positive youth development, and the role of meaningfulness of engagement. J. Posit. Psychol. **6**(1), 57–74 (2011)
10. Burtless, G. (ed.): Does Money Matter?: The Effect of School Resources on Student Achievement and Adult Success. Brookings Institution Press (2011)
11. Burton, J., Marshall, L.: Protective factors for youth considered at risk of criminal behaviour: Does participation in extracurricular activities help? Crim. Behav. Ment. Health. **15**(1), 46–64 (2005). https://doi.org/10.1002/cbm.36
12. Cheung, A.: "Friendship brought me into it, but commitment bought me into it": Civic engagement motivations among highly engaged Asian American youth. Teach. Coll. Rec. **123**(11), 102–124 (2021). https://doi.org/10.1177/01614681221087300
13. Delle-Vergini, S., Ally, M., Eacersall, D., Dann, C., Chakraborty, S.: Teaching project management to primary school children: Exploring the perspectives of project practitioners. Issues Educ. Res. **33**(1), 41–70 (2023)
14. Diener, E., Oishi, S., Lucas, R.E.: Personality, culture, and subjective well-being: Emotional and cognitive evaluations of life. Annu. Rev. Psychol. **54**, 403–425 (2003). https://doi.org/10.1146/annurev.psych.54.101601.145056
15. Edgerton, J.D., Roberts, L.W., von Below, S.: Education and quality of life. In: Handbook of social indicators and quality of life research, pp. 265–296 (2011)
16. Fitri, P.N., Setyowati, S.: The Implementation of SBM in Extracurricular Activities at Elementary school. In: 2nd International Conference on Education Innovation (ICEI 2018), pp. 343–346. Atlantis Press (2018, December)
17. Forneris, T., Camiré, M., Williamson, R.: Extracurricular activity participation and the acquisition of developmental assets: Differences between involved and noninvolved Canadian high school students. Appl. Dev. Sci. **19**(1), 47–55 (2015)
18. Fredricks, J., Eccles, J.: Is extracurricular participation associated with beneficial outcomes? Concurrent and longitudinal relations. Dev. Psychol. **42**(4), 698–713 (2006). https://doi.org/10.1037/0012-1649.42.4.698
19. Gajović, A., Bjelica, D., Pavlović, D., Vukmirović, D.: Educating youth on project sustainability: Project engagement and recognition of the green deal. Appl. Ecol. Environ. Res.: Int. Sci. J. **21**(4), 2969–2989 (2023). https://doi.org/10.15666/aeer/2104_29692989

20. Hancock, D., Dyk, P., Jones, K.: Adolescent involvement in extracurricular activities: Influences on leadership skills. J. Leadersh. Educ. **11**(1), 84–101 (2012) https://doi.org/10.12806/v11/i1/rf5
21. Hanushek, E.A., Woessmann, L.: School resources and student achievement: A review of cross-country economic research. In: Cognitive abilities and educational outcomes: A festschrift in honour of jan-eric gustafsson, pp. 149–171 (2017)
22. Heath, R., Anderson, C., Turner, A., Payne, C.: Extracurricular activities and disadvantaged youth: A complicated - but promising - story. Urban Educ. **57**(8), 1415–1449 (2018). https://doi.org/10.1177/0042085918805797
23. Huffman, C.L., Tallant, A.C., Young, S.C.: Preliminary impact of DegreePlus: An institutional program to provide transferable skill development through extracurricular activities. Student Success. **10**(1), 131–139 (2019)
24. Ilieva, S., Vitanova, V.: Career development expectations of young people. Knowl. Int. J. **34**(6), 1743–1748 (2019). https://doi.org/10.35120/kij34061743i
25. Janke, K.K., Nelson, M.H., Bzowyckyj, A.S., Fuentes, D.G., Rosenberg, E., DiCenzo, R.: Deliberate integration of student leadership development in doctor of pharmacy programs. Am. J. Pharm. Educ. **80**(1), 2 (2016)
26. Kramer, J., Barth, Y., Curtis, K., Livingston, K., O'Neil, M., Smith, Z., Wolfe, A.: Involving youth with disabilities in the development and evaluation of a new advocacy training: Project team. Disabil. Rehabil. **35**(7), 614–622 (2012). https://doi.org/10.3109/09638288.2012.705218
27. Mah, D.K., Ifenthaler, D.: What do first-year students need? Digital badges for academic support to enhance student retention. J. Appl. Res. High. Educ. **12**(1), 86–96 (2019)
28. Martinez, A., Coker, C., McMahon, S.D., Cohen, J., Thapa, A.: Involvement in extracurricular activities: Identifying differences in perceptions of school climate. Educ. Dev. Psychol. **33**(1), 70–84 (2016)
29. Medina, A., Hernández, J.C., Muñoz-Cerón, E., Rus-Casas, C.: Identification of educational models that encourage business participation in higher education institutions. Sustainability. **12**(20), 8421 (2020)
30. Meltzer, A., Powell, A., Saunders, I.: Pathways to educational engagement: an exploratory study of outcomes from an Australian school-based youth mentoring program. J. Youth Stud. **23**(5), 545–560 (2019). https://doi.org/10.1080/13676261.2019.1634800
31. Mihić, M., Bjelica, D.: Mladi u Srbiji: obrazovni, ekonomski i socijalni aspekti. Univerzitet Beograd, Fakultet organizacionih nauka (2024)
32. Mijacic, D., Paunović, B.: Regional disparities in Serbia. Ekon. Preduz. **59**(7-8), 379–389 (2011). https://doi.org/10.5937/ekopre1108379m
33. Mikovic, R., Petrovic, D., Mihic, M., Obradovic, V., Todorovic, M.: Examining the relationship between social capital and knowledge usage in the nonprofit industry. Knowl. Manag. Res. Pract. **17**(3), 328–339 (2019)
34. Noor, A.M., Priyalatha, G., Isa, N.J.M.: Personal growth initiative in relation to life satisfaction among university students: Inisiatif pertumbuhan personal dan perkaitannya dengan kepuasan hidup dalam kalangan pelajar universiti. Educatum J. Soc. Sci. **6**(2), 42–49 (2020)
35. Novytskyi, Y.: Modern: problems of developing a healthy lifestyle among young people. Sci. J. Natl. Pedagog. Dragomanov Univ. Ser. 15 Sci. Pedagog. Prob. Phys. Cult. (Phys. Cult. Sports). **12**(172), 9–12 (2023). https://doi.org/10.31392/udu-nc.series15.2023.12(172).01
36. Ozer, E., Douglas, L.: The impact of participatory research on urban teens: an experimental evaluation. Am. J. Community Psychol. **51**(1-2), 66–75 (2012). https://doi.org/10.1007/s10464-012-9546-2
37. Pavlović, D., Domazet, I., Lazić, M.: The impact of education on the youth labour market in Serbia. Int. Lett. Soc. Humanist. Sci. **83**, 11–18 (2018) https://doi.org/10.18052/www.scipress.com/ilshs.83.11

38. Paek, S., Leong, P., Johnson, P., Moore, C.: Is GPA enough? A platform for promoting computer science undergraduates' pursuit of career-related extracurricular activities. Int. J. Technol. Educ. Sci. **5**(1), 1–16 (2021)
39. Pham, Y.K., Murray, C.: Social relationships among adolescents with disabilities: Unique and cumulative associations with adjustment. Except. Child. **82**(2), 234–250 (2016)
40. Penn-Edwards, S., Donnison, S.: Engaging with higher education academic support: A first year student teacher transition model. Eur. J. Educ. **46**(4), 566–580 (2011)
41. Peréa, F., Sayles, N., Reich, A., Koomas, A., McMann, H., Martínez, L.: Mejorando nuestras oportunidades: engaging urban youth in environmental health assessment and advocacy to improve health and outdoor play spaces. Int. J. Environ. Res. Public Health. **16**(4), 571 (2019)
42. Petrović, D.: Implementacija agilnog upravljanja i poslovna agilnost. Tehnika. **77**(5), 485–491 (2022)
43. Powdthavee, N., Lekfuangfu, W.N., Wooden, M.: What's the good of education on our overall quality of life? A simultaneous equation model of education and life satisfaction for Australia. Int. J. Environ. Res. Public Health. **54**, 10–21 (2015)
44. Powers, J., Tiffany, J.: Engaging youth in participatory research and evaluation. J. Public Health Manag. Pract. **12**, S79–S87 (2006). https://doi.org/10.1097/00124784-200611001-00015
45. Pukkinen, T., Hytti, U., Heinonen, J., Stenholm, P.: Curricular and extracurricular entrepreneurial activities supporting entrepreneurial self-efficacy and desirability of rural youth. Entrep. Educ. Pedagog. **7**(3), 315–346 (2023). https://doi.org/10.1177/25151274231190806
46. Rais, S., Koh, D., Yaccob, Y.: Understanding factors of involvement and motives of youth in volunteerism. Int. J. Acad. Res. Bus. Soc. Sci. **11**(12) (2021). https://doi.org/10.6007/ijarbss/v11-i12/11752
47. Rošulj, D., Petrović, D.Č., Arsić, S.M.: Knowledge Management in Serbian SMEs: Key Factors of Influence on Internal and External Business Performances. Sustainability. **16**(2), 797 (2024)
48. Rubio, L., Mireles, C., Jones, Q., Mayse, M.: Identifying issues surrounding first generation students. Am. J. Undergrad. Res. **14**(1) (2017)
49. Schelven, F., Boeije, H., Inhulsen, M., Sattoe, J., Rademakers, J.: "We know what we are talking about": experiences of young people with a chronic condition involved in a participatory youth panel and their perceived impact. Child Care Pract. **27**(2), 191–207 (2019). https://doi.org/10.1080/13575279.2019.1680529
50. Schwartz, K., Cappella, E., Seidman, E.: Extracurricular participation and course performance in the middle grades: a study of low-income, urban youth. Am. J. Community Psychol. **56**(3-4), 307–320 (2015). https://doi.org/10.1007/s10464-015-9752-9
51. Slavinski, T., Bjelica, D., Pavlović, D., Vukmirović, V.: Academic performance and physical activities as positive factors for life satisfaction among university students. Sustainability. **13**(2), 497 (2021). https://doi.org/10.3390/su13020497
52. Smith, L., Baranowski, K., Abdel-Salam, L., McGinley, M.: Youth participatory action research: agency and unsilence as anti-classist practice. Transl. Issues Psychol. Sci. **4**(2), 176–186 (2018). https://doi.org/10.1037/tps0000155
53. Stanimirovic, P., Borozan, T., Radojicic, M., Djordjevic-Tomic, A.: Project Management Software Tools–One Step Closer to UN Sustainable Development Goals. In: 2023 3rd International Conference on Electrical, Computer, Communications and Mechatronics Engineering (ICECCME), pp. 1–5. IEEE (2023) ISBN: 979-8-3503-2297-2
54. Suraswadi, P.: The roles of actor/teacher in creating youth community theatre: empowering and re-connecting youths with their community in a partnership school project, Pichit Province. Thail. Parichart J. **36**(4), 161–175 (2023) https://doi.org/10.55164/pactj.v36i4.264525

55. Todorović, M.L., Petrović, D.Č., Mihić, M.M., Obradović, V.L., Bushuyev, S.D.: Project success analysis framework: A knowledge-based approach in project management. Int. J. Proj. Manag. **33**(4), 772–783 (2015)
56. Utomo, M.F.P., Dhyani, I.A.D., Yusari, I.G.A.A.A., Aryadi, I.P.H., Darmayanti, N.P.D.U., Ganesha, I.G.H.: Correlation between extracurricular activities and academic performances among preclinical medical students in Udayana University, Bali, Indonesia. Intisari Sains Med. **10**(3) (2019)
57. Wagner, T., Compton, R.A.: Creating Innovators: The Making of Young People who will Change the World. Simon and Schuster (2012)
58. Youniss, J., Bales, S., Christmas-Best, V., Diversi, M., McLaughlin, M., Silbereisen, R.: Youth civic engagement in the twenty-first century. J. Res. Adolesc. **12**(1), 121–148 (2002)

Project Management Maturity Evaluation: A Case Study from the Transport and Logistics Industry

Katarina Petrović(✉), Zorica Mitrović, and Tea Borozan

University of Belgrade, Faculty of Organizational Sciences, Jove Ilića 154, Belgrade, Serbia
katarina.petrovic@fon.bg.ac.rs

Abstract. An increasing number of companies are setting up Project Management Offices (PMOs) to benefit from project portfolio management and enhance their overall competitiveness in the market. However, it is never ending process of encountering challenges and identifying areas for improvement. This paper aims to apply Project Management Maturity Models (PMMMs) as a tool to assess the level of maturity in portfolio management in different project management knowledge areas. Specifically, this paper examines a selected case study from the Transport & Logistics industry, utilizing findings to propose strategies for improving process performances. Findings show that there are valuable benefits in setting up of PMO in the Transport & Logistics industry and that the PMMM^{sm} model is applicable in this specific industry. Finally, the paper enriches the existing literature in the field, but also provides strategies for practitioners looking to improve project performance through PMOs, especially in the Transport & Logistics industry.

Keywords: Project Management Office · Project Management Maturity Model · Transport & Logistics

1 Introduction

Today's fast-paced and highly demanding business environment continuously compels organizations to search for systems and solutions that will facilitate their ability to achieve or maintain competitive advantage [1]. Organizations adopt and enforce different structures, such as Project Management Offices (PMOs), to enhance execution efficiency and overall project success [2]. Project Management Offices can assist organizations in defining performance measures and different factors influencing project success [3], thereby enabling identification of challenges, issues and internal limitations and further facilitate formulation of rectifying activities [4]. PMOs can have a substantial contribution to project outcomes, which is supported by the fact that project quality can be augmented by approximately 25% [5]. Additionally, according to one research, it has been confirmed that three out of four PMOs have a highly beneficial impact on **project management** efficiency [6].

V. Štavljanin et al. (Eds.): SymOrg 2024, LNNS 1681, pp. 182–194, 2025.
https://doi.org/10.1007/978-3-032-08090-5_13

The attainment of performance enhancement is contingent upon awareness of organizational weaknesses. Project Management Maturity Model (PMMM) represents a valuable tool in this regard, as they support assessment and appraisal of specified maturity level of project management competences that have been developed over a certain period of time [7]. Research conducted by Gareeb and Rwelamila [8] reveals existence of large number of PMMMs, identifying up to 60 different models, by employing extensive literature review methodology for the period from 2000 to 2019. Nevertheless, it is necessary to acknowledge that these PMMMs are not universally applicable across all industries [9]. Based on organizational size, research shows that PMOs have been adopted by both medium and large organizations [10].

A growing number of various industries are acknowledging the importance and value of project management, but its role is being especially emphasized in the **Transport & Logistics** sector [11]. The underlying cause is in the rising complexity and the necessity for higher efficiency of project execution. Valued at $8.4 trillion during 2021, the global logistics market is being forecasted to exceed $13.7 trillion by 2027, which emphasizes the industry's expansion and the indispensable role of project management in facilitating this growth [12].

Research has indicated that the application of PMMM is predominantly concentrated in the information technology industry, while their utilization in all other sectors is less prevalent [13]. Despite the crucial role of projects within the Transport & Logistics industry, there remains a gap in comprehensive understanding of potential advantages and benefits that PMO can offer [14]. The literature review highlights deficiency of studies and research required to comprehend and perceive expected benefits and potential improvements associated with PMO implementation and the selection of suitable PMMM for the fast-growing Transport & Logistics sector.

The study is designed to examine the two research questions. First, do Transport & Logistics companies need a PMO and how should it be implemented. Second, which PMMM is appropriate and applicable for Transport & Logistics companies.

To answer this question the paper is structured into five sections. The literature review summarizes the research concepts applied in the study, as well as previous findings relevant to the study. The research design section offers an overview of the selected company and the methodological approach. The results and discussion section gives an insight into gathered data and provides an interpretation of the results. The conclusion highlights the principal findings and the paper's contribution to the topic. The final section addresses study limitations and offers recommendations for further research.

2 Literature Review

2.1 Project Management Office

Project Management Institute describes **Project Management Office (PMO)** as "organizational structure that standardizes the project-related governance processes and facilitates the sharing of resources, methodologies, tools, and techniques" [15]. The PMO aims to provide comprehensive support to project managers and their teams, as well as to the functional areas within the company, with the objective of emphasizing organizational overall effectiveness and efficiency [16]. PMO represents the joining point between organization and its projects, and an entity which provides possibilities for advancement and improvement of project management expertise [17].

The Project Management Institute [15] classifies PMOs into three primary types: supportive, controlling and directive. The supportive PMO functions in a consultative role, without the power or authority to issue directives. This PMO is responsible for development, securement and maintenance of standardized project templates, as well as assuring availability of required data and endorsing best practices. It is important to note that supportive PMO operates with the lowest level of authority over project teams. In addition to consultative role, the controlling PMO type has certain authority to oversee application of standardized methods and project templates. As the type of project management office with the highest level of authority, the directive PMO is entrusted with direct oversight and entirely accountable for overall project success [15].

Organizations need to perceive PMO as a dynamic component and structure, rather than a rigid framework, as it contributes to addressing ongoing challenges and issues that companies face due to fast-paced environment [18]. For a PMO to reach its full potential, the organization needs to consider that it undergoes continuous development and adapts to the context in which it operates [19]. The PMO is an essential factor for knowledge management and standardization of processes, thus aiming to prevent recurrence of past mistakes within the organization, despite the chosen PMO type. A study conducted at the institutional level revealed that institutions in the Republic of Serbia demonstrate a certain awareness level about knowledge management, but what is lacking is the formalization of accountability for managing it [20], which potentially indicates that the private sector may have similar implications and eventually fail to apply PMO for this purpose. In recent years a growing number of organizations decided to implement Project Management Office into their structure, with some companies even deciding to incorporate multiple PMOs [21]. Nevertheless, selecting and establishing a suitable, long-lasting PMO is a complex process, therefore assessment and clear understanding of organizations project management maturity signifies the initial step toward ensuring PMO success.

2.2 Project Management Maturity Model

The **Project Management Maturity Model (PMMM)** can be defined as a tool which provides a structured framework designed to evaluate organizational project management expertise and competences. PMMMs help organizations assess and identify their **maturity level**, highlight deficiencies and strengths, and guide further formulation of improvement strategies aimed at enhancing project management performance [4]. In this context, the term "maturity" denotes successful project delivery enabled through utilization of standardized procedures and continuous project execution [22]. PMMM provides a comparative analysis between organizational processes and best practices which have been adopted [23]. Project management maturity assessment implies inclusion of company's project-oriented nature and specific factors pertinent to its industry [24].

Through the literature review a range of models were identified. One study, which encompassed literature review from 2000 to 2019, revealed 60 various project management **maturity models**, among which are Prado's P2M3, CMMI, Kerzner's KPMMM, IPMA Delta, MMM, NPM3, OPM3, P3M3, SPM3 etc. [8].

Research performed by Simagunsong and Da Silva [25] aimed to identify the most frequently adopted project management maturity models across different industries, with a particular emphasis on models such as ProMMM, CMMI, PM2, PMMM^{sm}. The research results revealed that PM2, ProMMM and PMMM^{sm} models are appropriate for project-driven organizations. Furthermore, CMMI was described as a more suitable alternative for the software industry. While ProMMM has four maturity levels, comparative analysis highlighted both PM2 and PMMM^{sm} include five levels of maturity. Moreover, findings indicated that PMMM^{sm} encompasses both project and program management processes, while PM2 covers only project management processes [26, 27]. Considering the previously mentioned findings, this research paper seeks to assess the relevance and suitability of PMMM^{sm} regarding rapidly evolving and dynamic organizations within the Transport & Logistics industry.

2.3 PM Solutions Project Management Maturity Model

PM Solutions PMMM (PMMM^{sm}) represents a model that integrates five maturity levels with PMI's project management **knowledge areas** [28]. It provides a comprehensive framework for evaluation of project management maturity across various organizational areas. Each of the levels reflects a distinct degree of standardization and the use of established patterns, as presented in figure below [29, 30] (Fig. 1).

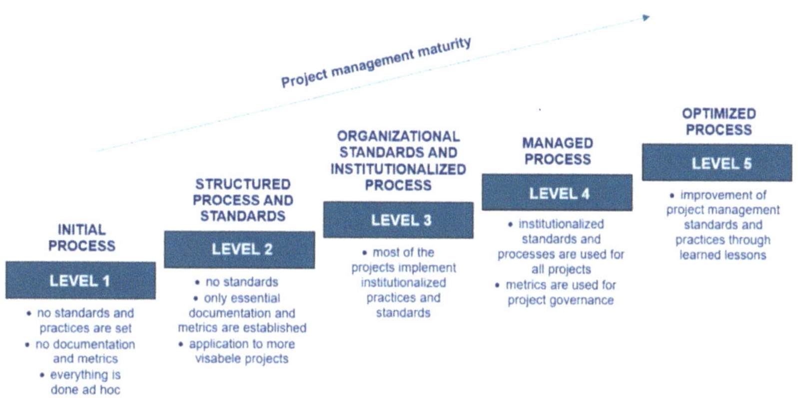

Fig. 1. Levels of PMMM^{sm}

The underlying reason for models' widespread utilization resides in its comprehensive framework which covers knowledge areas defined by PMI's Project Management Body of Knowledge (PMBOK). Pennypacker & Grant [31] based their study on this model, where they aimed to identify project management maturity level by including 123 respondents, across different industries, in the research. Miklosik [32] used PMMM^{sm}

as a foundation for evaluating project management maturity in 25 information and communications technology (ICT) companies in Slovakia. The model was also applied in a study by Göçmen [33] as a tool for analyzing maturity level of a logistics company, with a focus on sustainable project management.

3 Research Design

The purpose of this study is to examine the **implementation** of Project Management Office within a Transport & Logistics company through the lens of a specific Project Management Maturity Model. The selected company, recognized as a regional leader, has a project management office managing over sixty projects annually. Its exceptional achievements in business innovation and operational efficiency, makes it an exemplary case for analyzing the effects of PMO practices. The company's PMO presents a comprehensive setting for determining maturity and effectiveness of project management practices. The broad range of projects across various transport and logistics aspects provides a thorough examination of the PMOs performance and capabilities across diverse project types.

The research adopted a **case study** methodology to utterly inspect the research subject from both theoretical and practical aspects [34]. The case study used a questionnaire that was developed based on PMMM^{sm} and utilized a mixed-methods approach, combining both qualitative and quantitative data collection. It is important to note that a Likert scale, ranging from 1 (one) to 5 (five) was used for the collection of quantitative data. This integrative approach allowed detailed understanding of research subjects, obtaining not only quantitative data but also valuable qualitative opinions, perspectives and other insights.

The questionnaire was divided into five sections. The first section captured general information and analyzed the organizational approach to project management, specifically whether projects are being approached separately or within a larger strategic context. The second section aimed to identify the fundamental project management methodology adopted by the chosen company. The third section of the questionnaire evaluated the project management maturity level across six essential areas: integration, scope, time, cost, quality, and resources. The fourth section assessed the current implementation status of the existing project management office and conducted an in-depth analysis of its functionalities. The final section of the questionnaire reviewed adoption and use of project management software solutions.

The analysis process encompassed three key stages: data interpretation, member checking and final report preparation. Initially, the data were interpreted and rigorously scrutinized in the context of defined research objectives and theoretical viewpoint. Next, the findings were subjected to member checking, where representatives from the company reviewed the results to verify their accuracy and alignment with their perspectives and experiences. The final stage involved synthesizing the results and presenting them in a well-structured and systematic manner, which is addressed in the forthcoming section.

4 Research Results and Discussion

The data acquired from the questionnaire, indicates that the company manages an average of 60 projects annually, which are categorized as moderately complex, whereas financial value of the projects varies widely, ranging from several hundred to several million dollars. They are initiated either internally, for development purposes, or externally, to address business and operational requirements and needs. The average length of a project is approximately six months. It is important to highlight that the projects are subject to an in-depth analysis within the portfolio perspective, followed by a prioritization process that adheres to criteria defined and established by the company's project management office.

The chosen company applies both traditional and agile approaches. Consequently, project scope is defined either at the initial stage or later throughout the project life cycle. The agile methodology is particularly suited for projects where client requirements are unstable and subject to frequent alterations [35], thereby encouraging ongoing stakeholder engagement [36]. In contrast, the traditional methodology involves executing each project phase only once. Project success is determined based on accomplishment of project goals and degree of attained client satisfaction [37].

The project management maturity levels for the selected Transport & Logistics company, as presented in the figure below (Fig. 2), are the following: integration management – level 1, scope management – level 1, time management – level 4, cost management – level 1, quality management – level 1, and human resource management – level 3. Elaboration on the identified project management maturity levels will be further discussed.

Fig. 2. Project management maturity levels according to the knowledge areas

The current maturity of integration management is at level 1, reflecting the usage of ad hoc approach [29]. Decision-making is not supported by any performance data and metrics and there is a lack of institutionalized processes and standardized practices [38]. Moreover, the development of fundamental documentation and performance metrics is still unresolved.

The project scope management includes processes that ensure the project is executed according to defined plans and set objectives [39]. These processes are structured and organized into six main steps: conceptual development, scope statement, authorization of work, scope reporting, system monitoring and control, and project completion [39]. In the given company, scope management is at level 1, indicating that only general documentation has been structured, as a guideline for defined requirements. However, poor scope management reflects a lack of thorough upfront definition and planning, rather than simply inadequate documentation of requirements [40]. While management is acquainted with key milestones, stakeholder involvement is minimal, and they do not actively contribute to scope formulation process [29]. Neither of scope management methods or practices have been formalized or applied across undertaken projects. Project comparison and systematic analysis are not being conducted.

Time management is currently at level 4, representing the highest maturity degree which company has achieved compared to all project management knowledge areas. It reflects consistent application of established and standardized time management practices and processes to projects [30]. The use of historical data supports the creation of performance projections and predictions [41]. Decision-making processes are driven by already established and adopted metrics. Nonetheless, there is still a lack of continuous enhancement and improvement of practices based on past outcomes and project experience.

Insufficient project cost management can lead to budget overruns, delays, and failure to achieve project goals [42]. Cost management in this company is assessed at level 1, which shows reliance on ad hoc and informal practices due to unexecuted standardization of practices and processes [29]. There is no systematic approach to cost tracking or cost estimating, as well as metrics, which leads to decisions that lack metrics-based foundation. It is essential to carefully plan, monitor, and control costs at various stages of the project by utilizing modern tools and resource management methods, such as Lean Six Sigma, Total Cost Management, Activity-Based Costing, Earned Value Method, and Target Costing [42]. Additionally, leveraging modern project digitalization software like Microsoft Project, Primavera P6, Asana, and Trello [43] enhances the ability to effectively control and track project costs.

In today's competitive business environment, quality management is crucial in all aspects of a project. Implementing a specific standard of quality within an organization helps ensure customer satisfaction by delivering a high-quality product [44]. With quality management at level 1, it indicates that defined quality criteria and standards were either not established before starting the project or were insufficiently documented. Additionally, quality assurance reviews and reports may not have been thoroughly considered [44].

At level 3, as the second most developed project management knowledge area in this company, human resource management signifies established resource tracking methods and processes being applied to most of the projects [30]. Although management considers the professional development and overall career path, detailed evaluation of overall team performance is absent. Despite the achieved maturity level, the company still does not create any resource projections needed for successful resource allocation and planning.

PMO is responsible for defining tools, processes, patterns and methods, as well as monitoring their implementation [45]. In this company, PMO lacks the authority to manage projects or make strategic decisions, with its role confined to determining whether an initiative qualifies as a project. However, all project executions must be approved by higher level management. The PMO ensures access to necessary information and tools, while also overseeing the application of established practices and patterns. The existing project management office lacks decision-making authority in critical situations. From the software tool implementation perspective, it has been noted that a software tool is utilized for portfolio management. Project team members, who report to both PMO and functional managers, demonstrate a low level of awareness and view the established standards and practices as unnecessary for customer value delivery. This can be attributed to insufficient project management skills and experience. It is increasingly clear that success as a project manager cannot be achieved solely through technical skills [46]. Project managers working within a matrix organizational structure and leading large-scale initiatives face a communication dynamic unlike any other management role. Strong interpersonal or soft skills are essential for achieving success in this position [47]. Motivation of project team influences performance of a PMO, therefore it is important for project managers to enhance team motivation by developing deeper understanding of benefits that established procedures and standards can have. Various approaches, such as simulation games, have been proven to be effective for enhancing individual motivation [48].

The company appears to prioritize project portfolio management in order to establish the most effective combination of projects and employs hybrid project management methodology. The company acknowledges that the strategic benefits of portfolio management surpass those achieved through management of individual projects. It can be noted that the company uses a system-thinking approach that supports interests and requirements governance from different stakeholders, which contributes to project success enhancement [49].

According to the PMMM^{sm} model, the company shows a generally low project management maturity, with the exception of time management and human resource management, which are areas of higher priority and greater emphasis. The advanced maturity in human resource management is likely a response to labor shortages and turnover challenges in the Serbian logistics and supply chain sector [50]. While the company focuses on improvement of time management project practices and patterns as a key point for the logistics industry [51], it still demonstrates a lower level of maturity in terms of cost management.

The company has opted for a controlling project management office to effectively manage and oversee a substantial volume of external client projects and internal development projects. To bolster this initiative, the company implemented a project management software solution. The software's intuitive interface and extensive functionality were highlighted as significant benefits, allowing employees to manage their work and monitor project progress throughout a unified platform. The user-friendly design has notably facilitated the transition and adoption process for employees in the company [52]. Furthermore, the company is considering integration of other software solutions to enhance and improve its portfolio management capabilities.

5 Limitations and Future Directions

The primary limitation of this research paper is the small number of companies included in the case study. Future research should encompass a wider range of companies on an international level. Another constraint is the number of examined knowledge areas, which was limited to six areas, based on the initial $\mathrm{PMMM^{sm}}$ concept, but future studies should include and incorporate all existing knowledge areas, according to the Project Management Institute. Expanding the research to explore and validate multiple Project Management Maturity Models could provide deeper insights and findings.

It should be noted that the PMO contribution can provide significant benefits for numerous industries, including those of public importance, such as education [53], infrastructure [54], healthcare [55] and home care [56], energy [57], etc., which could serve as a guideline for expanding future research and examining the application and contribution of PMO in sectors other than Transport & Logistics.

6 Conclusion

The study was conducted to explore the following research questions: (1) Did Transport & Logistics companies require a PMO, and how should it have been implemented? (2) Which PMMM was most suitable and applicable for Transport & Logistics companies? The findings indicate that Transport & Logistics companies must establish a project management office in order to ensure successful management of such large number of projects. Successful implementation of PMO is contingent upon thorough understanding of the company's basis and fundamental characteristics. The Transport & Logistics companies usually handle two project types: those aimed at external clients and those dedicated to a company's internal needs. Simply put, these companies undertake both core business projects and projects focused on technological development. Although these two types of projects are equally important for overall business success, their natures are inherently different. Therefore, they cannot be handled in the same way and require methods, processes and patterns customized and tailored to their specific characteristics. To properly account for those differences, this research paper recommends the establishment of two separate Project Management Offices in Transport & Logistic companies. One office would be responsible for managing core business projects, while the other would focus on the organization's technological progress and innovation. Furthermore, each project management office should define and standardize appropriate patterns for each project type.

The findings for the second research question indicate that the $\mathrm{PMMM^{sm}}$ model is highly effective for **maturity evaluation** of these organizations. This model goes beyond just overviews, it provides detailed and comprehensive analysis of companies' maturity level from different aspects. In such a technologically demanding and rapidly changing industry, it is crucial to pinpoint areas that lack development and take necessary actions, where the direction of these steps is easily determined thanks to the $\mathrm{PMMM^{sm}}$.

This study revealed that adoption of Project Management Office could contribute to simplification of project management processes in the Transport & Logistics industry. Additionally, the $\mathrm{PMMM^{sm}}$ was identified as the most appropriate and beneficial model for this industry. The findings and results of this research paper provide a valuable guidance for project and strategic managers in the Transport & Logistics sector, assisting them in potential organizational structure adjustments and decision-making processes to improve project success, ultimately resulting in enhanced business outcomes.

References

1. Ferreira, J.J., Teixeira, A.A.C.: Open innovation and knowledge for fostering business ecosystems. J. Innov. Knowl. **4**(4), 253–255 (2019). https://doi.org/10.1016/j.jik.2018.10.002
2. Pellegrinelli, S., Garagna, L.: Towards a conceptualisation of PMOs as agents and subjects of change and renewal. Int. J. Proj. Manag. **27**(7), 649–656 (2009). https://doi.org/10.1016/j.ijproman.2008.12.001
3. Todorović, M., Mitrović, Z., Bjelica, D.: Measuring project success in project-oriented organizations. Manag. - J. Sustain. Bus. Manag. Solut. Emerg. Econ. (former.: Manag. – J. Theory Pract. Manag.). **68**, 41–48 (2013) (ISSN 0354-8635)
4. Kostalova, J., Tetrevova, L.: Proposal of project management methods and tools oriented maturity model. Rev. Gest. Proj. **9**(1), 01–23 (2018). https://doi.org/10.5585/gep.v9i1.595
5. De Miranda e Silva Correia, C., Moreira, C.M., Maia Muniz, R.: A importância da implantação do Escritório de Projetos: Estudo de caso de uma organização de médio porte. Sist. Gest. **13**(2), 141–148 (2018). https://doi.org/10.20985/1980-5160.2018.v13n2.1207
6. Ichsan, M., Sadeli, J., Jerahmeel, G., Yesica, Y.: The role of project management office (PMO) manager: A qualitative case study in Indonesia. Cogent Bus. Manag. **10**(2), 2210359 (2023). https://doi.org/10.1080/23311975.2023.2210359
7. Christoph, A.J., Konrad, S.: Project complexity as an influence factor on the balance of costs and benefits in project management maturity modeling. Procedia: Soc. Behav. Sci. **119**, 162–171 (2014). https://doi.org/10.1016/j.sbspro.2014.03.020
8. Gareeb, N., Rwelamila, P.D.: Rethinking project management maturity models for the South African power sector. Afr. J. Sci. Technol. Innov. Dev. **14**(3), 759–773 (2022). https://doi.org/10.1080/20421338.2021.1899558
9. Grobler, J., Steyn, H.: Project management maturity models: Does one size fit all? In: PMSA International Conference: Growth and Collaboration for a Project Management Profession. Project Management South Africa (PMSA), Johannesburg (2006)
10. PM Solutions: The State of the Project Management Office (PMO) 2022. PM Solutions, Inc., Chadds Ford, Pennsylvania (2022)
11. Hartel, D.H.: Classification and basics of project management in logistics and SCM. In: Project Management in Logistics and Supply Chain Management. Springer series in supply chain management, pp. 3–41. Springer Fachmedien Wiesbaden, Wiesbaden, Germany (2022). https://doi.org/10.1007/978-3-658-35882-2_1
12. Placek, M.: Third-party logistics (3PL) industry worldwide - statistics & facts. https://www.statista.com/topics/7122/third-party-logistics-3pl-industry-worldwide/#topicOverview (2024). Accessed 28 Apr 2024.
13. De Souza, T.F., Gomes, C.F.S.: Assessment of maturity in project management: A bibliometric study of main models. Procedia Comput. Sci. **55**, 92–101 (2015). https://doi.org/10.1016/j.procs.2015.07.012
14. Saedi, S.: An Introduction to Project Divisions: Why 3PL Providers Need Project Management Offices. https://www.linkedin.com/pulse/introduction-project-divisions-why-3pl-providers-need-soheil-saedi (2024). Accessed 05 May 2024.
15. Project Management Institute: A Guide to the Project Management Body of Knowledge (PMBOK Guide), 6th edn. Project Management Institute, Inc., Newtown Square, PA (2017)

16. Rezende Oliveira, R., Cordeiro Martins, H.: Desempenho do Escritório de Gerenciamento de Projetos: Divergência de percepções entre a instrução acadêmica e a vivência profissional. Innovar. **30**(75), 119–134 (2020). https://doi.org/10.15446/innovar.v30n75.83261
17. Khoori, A., Hamid, S.: A Study on the project management practices for PMO and public sector—A systematic review. Eur. J. Mol. Clin. Med. **7**(3), 767–774 (2020)
18. Aubry, M., Hobbs, B., Müller, R., Blomquist, T.: Identifying forces driving pmo changes. Proj. Manag. J. **41**(4), 30–45 (2010). https://doi.org/10.1002/pmj.20191
19. Aubry, M., Hobbs, B., Thuillier, D.: Organisational project management: An historical approach to the study of PMOs. Int. J. Proj. Manag. **26**(1), 38–43 (2008). https://doi.org/10.1016/j.ijproman.2007.08.009
20. Mitrović, Z., Obradović, V., Suknović, M.: Knowledge management in the public sector: The case of Serbian local government. Serb. J. Manag. **13**(2), 293–309 (2018). https://doi.org/10.5937/sjm13-15037
21. Müller, R., Glückler, J., Aubry, M.: A relational typology of project management offices. Proj. Manag. J. **44**(1), 59–76 (2013). https://doi.org/10.1002/pmj.21321
22. Anantatmula, V.S., Rad, P.F.: Role of organizational project management maturity factors on project success. Eng. Manag. J. **30**(3), 165–178 (2018). https://doi.org/10.1080/10429247.2018.1458208
23. Silva, C., Ribeiro, P., Pinto, E.B., Monteiro, P.: Maturity model for collaborative r&d university-industry sustainable partnerships. Procedia Comput. Sci. **181**, 811–817 (2021). https://doi.org/10.1016/j.procs.2021.01.234
24. Kerzner, H.: Strategic Planning for Project Management Using a Project Management Maturity Model, 3rd edn. Wiley, Hoboken, New Jersey (2019)
25. Simangunsong, E., Da Silva, E.N.: Analyzing project management maturity level in indonesia. the south east asian. J. Manag. **7**(1), 1–11 (2013). https://doi.org/10.21002/seam.v7i1.1521
26. Nikolaenko, V., Sidorov, A.: Assessment of project management maturity models strengths and weaknesses. J. Risk Financ. Manag. **16**(2), 121 (2023). https://doi.org/10.3390/jrfm16020121
27. Fabbro, E., Tonchia, S.: Project management maturity models: Literature review and new developments. J. Mod. Proj. Manag. **8**(3), 31–48 (2021)
28. Domingues, L., Ribeiro, P.: Project management maturity models: Proposal of a framework for models comparison. Procedia Comput. Sci. **219**, 2011–2018 (2023). https://doi.org/10.1016/j.procs.2023.01.502
29. Miller, B.: The pathway to OPM3: A busy project manager's guide to advancing organizational maturity. In: PMI® Global Congress 2004—North America, Anaheim. Project Management Institute, Newtown Square, PA (2004)
30. Vergopia, C.: Project review maturity and project performance: An empirical case study. Doctoral Dissertation. University of Central Florida (2008)
31. Pennypacker, J.S., Grant, K.P.: Project management maturity: An industry benchmark. Proj. Manag. J. **34**(1), 4–11 (2003). https://doi.org/10.1177/875697280303400102
32. Miklosik, A.: Improving project management performance through capability maturity measurement. Procedia Econ. Finance. **30**, 522–530 (2015). https://doi.org/10.1016/S2212-5671(15)01264-2
33. Göçmen, E.: A maturity model for assessing sustainable project management knowledge areas: A case study within a logistics firm. J. Adv. Res. Nat. Appl. Sci. **7**(4), 536–555 (2021). https://doi.org/10.28979/jarnas.958605
34. Ebneyamini, S., Sadeghi Moghadam, M.R.: Toward developing a framework for conducting case study research. Int. J. Qual. Methods. **17**(1), 1–11 (2018). https://doi.org/10.1177/1609406918817954

35. Project Management Institute: A Guide to the Project Management Body of Knowledge (PMBOK Guide), 7th edn. Project Management Institute, Newtown Square, Pennsylvania (2021)
36. Mitrović, Z., Mihić, M.: Projektno upravljanje razvojem softvera - Agilni pristup. In: Primena specijalizovanih menadžment disciplina u unapređenju konkurentnosti srpske privrede, pp. 9–26. Fakultet organizacionih nauka, Beograd (2017)
37. Sastoque-Pinilla, L., Artelt, S., Burimova, A., Lopez de Lacalle, N., Toledo-Gandarias, N.: Project success criteria evaluation for a project-based organization and its stakeholders—A Q-Methodology approach. Appl. Sci. **12**(21), 11090 (2022). https://doi.org/10.3390/app122111090
38. Forslund, H., Jonsson, P.: Dyadic integration of the performance management process. Int. J. Phys. Distrib. Logist. Manag. **37**(7), 546–567 (2007). https://doi.org/10.1108/09600030710776473
39. Al-Rubaiei, Q.H.S., Nifa, F.A.A., Musa, S.: Project scope management through multiple perspectives: A critical review of concepts. AIP Conf. Proc. **2016**, 020025 (2018). https://doi.org/10.1063/1.5055427
40. Mochal, T., Mochal, J.: Use scope change management. In: Lessons in Project Management, pp. 45–48. Apress, Berkeley, CA (2011). https://doi.org/10.1007/978-1-4302-3835-5_12
41. Hwang, S., Liu, L.Y.: Proactive project control using productivity data and time series analysis. In: Computing in Civil Engineering (2005), pp. 1–11. American Society of Civil Engineers, Reston, VA (2005). https://doi.org/10.1061/40794(179)179
42. Povazhnyi, O., Latysheva, O., Moiseienko, K., Chupryna, Y.: Peculiarities of project cost management. Econ. Her. Donbas. **1**(71), 4–10 (2023). https://doi.org/10.12958/1817-3772-2023-1(71)-4-10
43. Stanimirović, P., Borozan, T., Radojičić, M., Tomić, A.Đ.: Project management software tools – one step closer to un sustainable development goals. In: 2023 3rd International Conference on Electrical, Computer, Communications and Mechatronics Engineering (ICECCME), pp. 1–5. IEEE, Tenerife, Canary Islands, Spain (2023). https://doi.org/10.1109/ICECCME57830.2023.10252628
44. Alharthi, G., Khayyat, M.: The role of quality management in it project management. SAR J. - Sci. Res. **5**(2), 105–110 (2022). https://doi.org/10.18421/SAR52-06
45. Spalek, S.: Analytical challenges of a modern PMO. In: Spalek, S. (ed.) Data Analytics in Project Management, pp. 43–62. Auerbach Publications, Boca Raton, Florida (2018). https://doi.org/10.1201/9780429434891-4
46. Gillard, S.: Soft skills and technical expertise of effective project managers. Issues Informing Sci. Inf. Technol. **6**, 723–729 (2009). https://doi.org/10.28945/1092
47. Martínez Oropesa, C.: The essential soft skills of project managers and project success. Sci. J. Appl. Soc. Clin. Sci. **4**(13), 2–8 (2024). https://doi.org/10.22533/at.ed.2164132420062
48. Borozan, T., Maričić, M., Bjelica, D.: Impact of business simulation games on skills of engineering undergraduates. Int. J. Eng. Educ. **40**(5), 1090–1100 (2024)
49. Mitrović, Z.M., Rakićević, A.M., Petrović, D.Č., Mihić, M.M., Rakićević, J.Đ., Jelisić, E.T.: Systems thinking in software projects-an artificial neural network approach. IEEE Access. **8**, 213619–213635 (2020). https://doi.org/10.1109/ACCESS.2020.3040169
50. Kilibarda, M., Pajić, V., Andrejić, M.: Human resources in logistics and supply chains: Current state and trends. Int. J. for Traffic and Transp. Eng. **9**(3), 270–279 (2019). https://doi.org/10.7708/ijtte.2019.9(3).01
51. Shamina, L., Borisova, I.: Time-based management concept on the basis of the organizational-managerial innovations in a logistic company. MATEC Web Conf. **239**, 03012 (2018). https://doi.org/10.1051/matecconf/201823903012

52. Yamada, M.: Smartsheet vs. Excel: Which is the Better Project Management Tool?. https://www.linkedin.com/pulse/smartsheet-vs-excel-which-better-project-management-yamada-cfp- (2024). Accessed 04 May 2024.
53. Kuzmenko, N., Kichuk, Y., Lesina, T., Levytska, L., Kostrytsia, N., Mazur, N.: Management of educational projects on the example of accreditation of educational programs. J. Curric. Teach. **11**, 264–274 (2022). https://doi.org/10.5430/jct.v11n1p264
54. Hordieieva, I., Omelyanenko, V., Krysovatyy, I., Oprysok, M., Kostin, I.: Project management in complex technical infrastructure projects: Challenges And Strategies. Ad alta: J. Interdiscip. Res. **14**, 88–93 (2024). https://doi.org/10.33543/1401398893
55. Mitrović, Z., Vujošević, M., Savić, G.: Data envelopment analysis for evaluating serbia's health care system. Manag. - J. Theory Pract. Manag. **75**, 39–46 (2015). https://doi.org/10.7595/management.fon.2015.0012
56. Mihić, M., Todorović, M., Obradović, V., Mitrović, Z.: Can we do better? Economic analysis of human resource investment to improve home care service for the elderly in Serbia. Clin. Interv. Aging. **11**(28), 85–96 (2016). https://doi.org/10.2147/CIA.S96778
57. Mihić, M., Petrović, D., Vučković, A., Obradović, V., Djurović, D.: Application and importance of cost-benefit analysis in energy efficiency projects implemented in public buildings: The case of Serbia. Therm. Sci. **16**(3), 915–929 (2012). https://doi.org/10.2298/TSCI110911090M

Implementation and Challenges of ISO 9001 Quality Management Systems in Local Self-Government: A Case Study of Serbian Municipalities

Sara Dimitrijević(✉), Jelena Ruso, and Ana Rakić

Faculty of Organizational Sciences, University of Belgrade, Jove Ilića 154, 11000 Belgrade, Serbia
sara.dimitrijevic@fon.bg.ac.rs

Abstract. This paper evaluates the implementation of **ISO 9001** quality management systems (QMS) in local self-government institutions in the Republic of Serbia, with a focus on its effectiveness in enhancing operational performance and service delivery. The research draws on a diagnostic model provided in Annex A of the ISO 18091 standard, specifically tailored for local governments, to assess the level of compliance across 39 performance indicators. The study highlights the critical importance of organizational capacity in implementing QMS in local administrations, which often operate with fewer resources compared to regional and national public bodies. Although ISO 9001 certification is widely recognized, its real value lies in effective application. This research suggests that integrating ISO 9001 with other management systems, such as ISO 14001 for environmental management, can bring greater benefits. Additionally, the paper calls for further reforms and training in local self-government to improve the consistency and reliability of public service delivery.

Keywords: Integrated Management System (IMS) · ISO 18091 · Local Self-Government

1 Introduction

Modern societies constantly adapt to changes in the lives and work of their citizens. In this context, the methods used to implement activities, and the role of public administration are of crucial importance to the state. Citizens are becoming more aware of their rights and the government's responsibilities, which is driving the transformation of public administration's role and the expansion of its activities to meet evolving societal needs and expectations (Gębczynska and Wolniak 2018).

In the current framework of public administration systems, the focus is on process efficiency and the outcomes of implemented activities. Among the many approaches to improving these processes, Ćwiklicki et al. (2021) highlight the ISO 9001 Quality Management System (QMS) as "one of the most recognized systems." Implementing

V. Štavljanin et al. (Eds.): SymOrg 2024, LNNS 1681, pp. 195–212, 2025.
https://doi.org/10.1007/978-3-032-08090-5_14

a QMS in line with ISO 9001 requirements involves significant effort and resource allocation. This is particularly relevant for local governments, which often operate with fewer resources compared to their regional and national counterparts. According to Ćwiklicki et al. (2021), concept of organizational resource capacity holds particular significance for local governments.

It's important to note that the factors supporting the implementation of ISO 9001 may vary across sectors (Almeida et al. 2018). Additionally, companies within the same industry may perceive the barriers to ISO 9001 certification differently (Sfakianaki and Kakouris 2018).

The objective of this work is to assess and identify the implementation of the QMS according to ISO 9001 requirements within local self-government institutions in the Republic of Serbia, specifically in Serbian municipalities. The study also presents the application of the ISO 18091 standard, which provides guidelines for ISO 9001 implementation in local self-government. The analysis was based on a questionnaire found in Annex A of ISO 18091, which serves as a diagnostic tool for self-evaluation. This tool enables the highest local authorities to assess performance across 39 indicators, with the questionnaire completed by employees from various municipalities in Serbia.

2 Integrated Management Systems

Modern market demands place the need on organizations to simultaneously implement various management systems to survive in the competitive environment and meet the requirements of a growing number of stakeholders. The number of management systems that are frequently applied is increasing over time. Therefore, organizations today should implement quality management systems, environmental management systems, occupational health and safety, information security, food safety, and similar systems to meet the needs and requirements of customers, employees, the state, and the environment. „Achieving goals is impossible without understanding and applying a quality management way of thinking" (Ruso et al. 2024).

Also, the standards requirements related to these management systems are often compatible. To make the process of implementing these systems more efficient, organizations can decide to integrate their management systems (Rebelo et al. 2014a). This way of implementing different management systems leads to the concept of Integrated Management Systems (IMS). There are different definitions of this term, but one way it was defined by Karapetrovic and Willborn (1998) is: "Integrated management systems represent a set of interrelated processes that share a set of people, information, materials, infrastructure, and financial resources to achieve composite objectives related to satisfying different stakeholders."

When an organization decides to do so, it achieves better efficiency in managing multiple management systems, as well as improving its external image and relationships with stakeholders (Talapatra et al. 2019).

When speaking about the integration of management systems, the most integrated are three systems: quality, environment, and occupational health and safety (BSI 2016; Muzaimi et al. 2018, Talapatra et al. 2019). Thus, when we talk about management system standards, ISO 9001, ISO 14001, and ISO 45001 standards can be integrated.

Such integration is possible because the requirements of these standards are compatible and similar.

The reasons why organizations decide on this integration can be that it brings numerous benefits and simplifications in their operations. Accordingly, many authors highlight various advantages of this integration, which are identified through the analysis of papers that mention the benefits of integrated management systems. The highlighted benefits are presented in Table 1.

Table 1. Benefits of Integrated Management Systems

No.	Benefits of IMS	References
1.	Promotes synergies	Paulikova et al. (2022), Ribeiro et al. (2017), Simon et al. (2013)
2.	Cost savings (cost benefit, saving financial sources)	Paulikova et al. (2022), Dahlin and Isaksson (2017), Ribeiro et al. (2017) Almeida et al. (2014), Simon et al. (2013)
3.	Improvements on internal efficiency and effectiveness	Paulikova et al. (2022), Ribeiro et al. (2017), Olaru et al. (2014)
4.	Avoid duplications	Dahlin and Isaksson (2017), Ribeiro et al. (2017)
5.	Improvement of organization image	Paulikova et al. (2022), Ispas and Mironeasa (2022), Ribeiro et al. (2017)
6.	Prominent position in the market (compititive in the market)	Šućurović et al. (2024), Almeida et al. (2014)
7.	Documents reduction	Dahlin and Isaksson (2017), Ribeiro et al. (2017)
8.	Effective management systems that organize, simplify (Standardization/ simplification/ organization)	Ribeiro et al. (2017)
9.	Customer's requirements (pressures from interested parties such as customers and competitors)	Ispas and Mironeasa (2022), Ribeiro et al. (2017)
10.	Internationalization	Šućurović et al. (2024), Ribeiro et al. (2017)
11.	Reduction of resources	Ispas and Mironeasa (2022), Ribeiro et al. (2017)

(continued)

Three levels of integration have been identified in the literature: no integration, where management systems are implemented separately; partial integration, which combines

Table 1. (*continued*)

No.	Benefits of IMS	References
12.	Efficient mss	Ribeiro et al. (2017)
13.	Increasingly competitive economies	Almeida et al. (2014)
14.	Quality products	Šućurović et al. (2024)
15.	New business opportunities	Paulikova et al. (2022)
16.	Improvement of organizational and social culture	Paulikova et al. (2022), Babeľová and Starecek (2021)
17.	Improvement of productivity and organization efficiency	Paulikova et al. (2022)
18.	Better communication between departments	Pinto et al. (2022), Dahlin and Isaksson (2017)
19.	Operational improvements	Pinto et al. (2022), Dahlin and Isaksson (2017)

audits but keeps documentation processes separate; and full integration, where all management system functions operate at an integrated level (Ikram et al. 2020). Telepatra et al. (2018) propose a more comprehensive model based on total quality management, leading to full integration. This approach fosters cross-functional competence among standards, adds value for both organizations and stakeholders, resolves conflicts between individual systems, and simplifies implementation. It is adaptable to organizations of various sizes, industries, and market positions, making it a practical solution for achieving system-wide synergy. Also, Barbosa et al. (2022) point out one more degree of integration - added integration, which includes objectives and documents.

The importance and relevance of this concept are further emphasized by the fact that the British Standards Institution (BSI) developed a publicly available specification (PAS – Publicly Available Specification) called BSI PAS 99 Integrated Management Systems. This document was developed in accordance with the ISO organization's guidelines for developing management system standards. A publicly available specification is "a document published by ISO or IEC in response to urgent market needs, which represents either a consensus within an organization outside ISO or IEC, or a consensus among experts within a working group" (ISS 2023). In addition, it is noted that publicly available specifications must comply with existing international standards, can coexist on the same subject, but are not recognized by organizations like CEN (European Committee for Standardization), CENELEC (European Committee for Electrotechnical Standardization), or ETSI (European Telecommunications Standards Institute) (ISS 2023). PAS 99 is the first specification for integrated management systems that provides a framework for implementing these systems. Some of the benefits of applying PAS 99 are that it allows organizations to meet all standard requirements with a single set of policies and procedures, enables audits of more than one system at the same time, saving money and resources, improves overall business efficiency by eliminating the need for task duplication, provides clearly defined roles and responsibilities that emphasize common goals, and enables continuous improvement of all management systems (BSI 2024).

3 Public Administration

The public administration of each country is characterized by a deep connection with the historical background of political systems, the culture and tradition of the people on one hand, and the further strategic orientations of each nation on the other. For these reasons, public administration is a very complex concept. Generally, the uniqueness of public administration for each country is reflected in its size, structuring model, scope of activities through institutions, and the manner of organization. "Public administration is the use of managerial, political and legal theories and processes to meet the legislative, executive and judicial mandates for government services and regulatory functions for society as a whole, or for specific segments", according to Ruso et al. (2013: 550).

The scientific study of public administration often emphasizes the unique features and characteristics of national administrative traditions. These traditions typically vary across several dimensions and are categorized into distinct typologies (Mavrot et al. 2021). In line with this, long-initiated globalization processes, changes from a linear economy to a circular economy, the needs of a growing population, and a series of other changes influence the way the world and countries are connected through policies and business cooperation on various levels. The dynamics of these activities, among other things, create challenges for public administration.

Public administration reforms through restructuring and reorganization, which began more than 40 years ago (Lapuente and Van de Walle 2020), have taken multiple directions of change in most countries. According to Lapuente and Van de Walle (2020,3), New Public Management is a broad and often ambiguous concept, with nearly every public sector reform in recent decades being labeled as such, even when it does not align with its fundamental principles.

There are several characteristic changes. The first is the abandonment of strict vertical hierarchy, i.e., the shift from centralization to decentralization. It should be noted that in this context, decentralization appears in three forms: administrative, fiscal, and political decentralization (Goel et al. 2017; European Commission 2017d; Dick-Sagoe 2020). The types of decentralization can act independently or in combination within reforms, depending on the activities.

According to Dick-Sagoe (2020:2), "The administrative part considers the organizational structure of local governments, while the political dimension involves politically appointed leaders heading local governance. The fiscal dimension focuses on financial arrangements and cost-sharing between local authorities and central government.".

Decentralization essentially involves the transfer of authority from central to local (self) governance. This approach also means a better understanding of public sector service users, i.e., addressing the specific needs of the population (Goel et al. 2017; European Commission 2017d).

There is extensive literature citing significant reasons for public administration reform towards decentralization. Some of these reasons include: local government being more mobile in obtaining information on the specific needs of the local population (Dick-Sagoe 2020; Ingrams et al. 2020; Pittaway and Montazemi 2020), the possibility of providing higher-quality services to the local population (Anttila and Jussila 2017; Bencsik et al. 2019), the ability to assess resource utilization from general to specific (Ćwiklicki et al. 2021), local governments are well-positioned to identify economically sustainable areas within their regions for taxation, facilitated by effective fiscal decentralization, according to Dick-Sagoe (2020), the possibility of better organizing and managing at the local level (Anttila and Jussila 2017; Almeida et al. 2018; Dick-Sagoe 2020; Tolo 2023), and others.

In these reforms, decentralization processes were accompanied by other processes and sub-processes, resulting in new or adapted forms of the public sector and public administration, which posed significant challenges for policymakers. Some of these processes were carried out horizontally, such as merging territorial units (so-called regionalization), while others involved separation (European Commission 2017d). On the other hand, the question arises vertically whether decisions are made closer to the center (national and supranational levels), i.e., centrifugal, or closer to the population, i.e., centripetal (European Commission 2017d). The latter represents the full application of the subsidiarity principle, which was incorporated into Article 5(3) of the Treaty of Lisbon, signed in 2007 (Treaty of Lisbon 2007).

A key factor in the application of multi-level governance (MLG) is the introduction and implementation of the subsidiarity principle. The Committee of Ministers of the member states, in recommendation CM/Rec (2022)10, describes multi-level governance as "a governance model that includes central, regional, and local governments, as well as civil society organizations. The ways in which it is organized can vary significantly from country to country. Ideally, it includes a bottom-up element and involves establishing participatory processes for co-creating policies, collaboration, and coordination among all relevant public bodies, at all levels of governance, and with all relevant stakeholders, in areas of shared competence or common interest."

The subsidiarity principle ensures that decisions are made as close as possible to the citizens. The principle of subsidiarity aims to ensure a certain level of autonomy for lower authorities in relation to higher ones, or for local governments relative to central governments. It establishes a division of powers across multiple levels of government, serving as the institutional foundation for federal states. Within the EU context, subsidiarity governs the use of the Union's non-exclusive powers, preventing intervention when issues can be effectively managed at local, regional, or national levels. The Union may exercise its powers only when member states are unable to achieve the proposed objectives effectively, and when action at the Union level offers added value. (European Parliament 2024).

Good governance must be based on appropriate principles and values. According to the OECD (2023): "Public governance principles are a comprehensive framework of standards that express the values and behaviors that citizens and businesses expect from modern public administration." The OECD (2023) also outlines six thematic areas that follow public governance principles, which are:

- strategy,
- policy development and coordination,

- public services and human resource management,
- oversight, accountability, and organization,
- digitization and service delivery, and
- public financial management.

Following these areas, the OECD (2023) outlines 32 principles of public governance. However, according to the European Commission (2017b:6), there are those that are recurrent in the administrations of EU member states, which are Legality, Integrity and Impartiality.

3.1 Local Self-Government

Several concepts are closely related to local self-government, yet it is important to distinguish between them. These concepts include public administration, the public sector, and local self-government. As Vukosavljević et al. (2021:41) state, "the public sector is the mechanism through which each state addresses key areas of collective work and activities for all its citizens." In modern states, citizens expect comprehensive services and protection, which public administration provides through various social and regulatory measures, managing public resources and overseeing private industries.

Local self-government, on the other hand, refers to "the right of citizens to manage public affairs that are of direct, common, and general interest to the local population, either directly or through freely elected representatives within local self-government units. It also includes the right and ability of local self-government bodies to, within the bounds of the law, manage affairs of public interest to the local population and within their jurisdiction" (Zakon o lokalnoj samoupravi 2021).

The internationalization of local self-government began in 1985 with the adoption of the European Charter of Local Self-Government by the Council of Europe, which came into effect in 1988. Since then, the Council has continually supported its development. An additional protocol (CETS No. 207) has further strengthened the charter by guaranteeing rights such as local referendums and civic participation in governance. The Committee of Ministers, in 2022, recommended a multi-level governance model that encourages collaboration between central, regional, and local governments, as well as civil society, fostering participatory policymaking and coordination (Council of Europe 1985).

According to Pittaway et al. (2020), local governments are in a prime position to integrate data from government, private entities, and citizens to create innovative public services. To make this possible, managers require training in leading the implementation of integrated enterprise systems. Andrews et al. (2020) emphasize the entrepreneurial role of local self-government bodies, highlighting initiatives like forming companies to provide public services, reflecting a variety of approaches seen globally. This illustrates the crucial role that local governments play in society, particularly in driving economic activity and providing services to citizens.

3.2 Local Self-Government in Republic of Serbia

In the late 18th and early 19th centuries, Serbia developed a system of local self-government, initially governed by village leaders known as princes. Over time, this system evolved, undergoing changes in both structure and funding. The Serbian Constitution of 1990 formally established the municipality as the fundamental unit of local governance. The 2006 Constitution further reinforced the citizens' right to local self-government, with specific laws outlining its organization and functions. The Ministry of State Administration and Local Self-Government, established in 2014, is responsible for overseeing local governance, including managing citizen registries and supporting local governments through capacity building and funding.

4 Quality Management in Local Self-Government

The introduction of quality management in public services aims to enhance organizational performance and improve national competitiveness (Lee Yong Wook et al. 2014). The public sector is expected to continuously improve the quality of its services to meet the needs and expectations of its primary users - citizens. As a result, the role and scope of public administration evolve in response to society's changing needs and demands (Gębczyńska and Wolniak 2018). A successful quality management system (QMS) could be significant for uprising organisational culture and managerial approaches (Milošević et al. 2021). Numerous methods, standards, and excellence models provide requirements and guidelines for implementing quality management systems. The ISO 9001 standard remains the most widely used method. According to the International Organization for Standardization (ISO 2022a), the number of ISO 9001 certificates reached 1,265,216 in 2022, making it the most certified standard globally. Of this total, 4,512 certificates were issued within the public administration sector.

"Standardisation is a key part of the microeconomic infrastructure with the benefits arising from the reduction of costs and the growth of quality", according to Trajković and Milošević (2018: 673). The complexity of applying this standard to all organizational processes is underscored by the existence of various standards with guidelines for specific types of organizations to implement ISO 9001 based on their activities. For public administration, ISO 18091:2022 – Quality management systems – Guidelines for implementation in local self-government is available.

Numerous authors have highlighted the benefits and importance of ISO 9001 for various local self-government organizations (Cho and Pyun 2022). However, it is crucial to emphasize that certification alone is not enough if the standard is not genuinely applied in practice. Mizuno (2002) examined the implementation of ISO 9001 in Japanese local governments and found that local governments can achieve similar benefits as private companies from applying the standard. However, authors like Löffler and Vintar (2004) and McAdam and Walker (2003) agree that ISO 9001 has a greater impact when integrated with other management systems, such as ISO 14001 – Environmental Management System.

The growing public awareness of the role of public administration has led local self-government organizations worldwide to focus on improving operational efficiency and implementing quality management systems.

5 Standard of Implementation in ISO 18091 – Example Local Self-Government

Organizations, regardless of their size or field of activity, frequently seek ISO 9001 certification to establish and regulate their quality management systems (QMS). This standard provides a framework for setting up a QMS and is widely regarded as a "best practice" benchmark. For those organizations aiming to enhance their performance further and achieve excellence, ISO 9004:2018 – Quality Management – Quality of an Organization – Guidelines for Achieving Sustainable Success, is available.

The ISO 18091:2022 standard serves to bridge the quality management concepts outlined in the ISO 9000 series with the specific practices and terminologies of local self-government. It aids local governments in implementing effective quality management systems (ISO 2022). These guidelines are aligned with the requirements of ISO 9001 and can be applied to all processes within local governments, covering strategic, tactical-management, and operational levels. This ensures the creation of a comprehensive quality management system that focuses on achieving the objectives of local administration (see Fig. 3).

By applying this standard and its guidelines, local governments can improve their operations and work towards excellence. However, as illustrated in Fig. 1, this standard alone does not guarantee excellence. To achieve that, local governments can adopt models like the European Model for Quality Management or the Malcolm Baldrige Excellence Model. Compliance with the ISO 9001 standard should not be viewed as the ultimate goal. Instead, once a local government reaches a level where it consistently provides products and services that meet community needs, it should consider using ISO 9004 and/or other excellence models to further enhance its overall effectiveness and efficiency (ISO 2022).

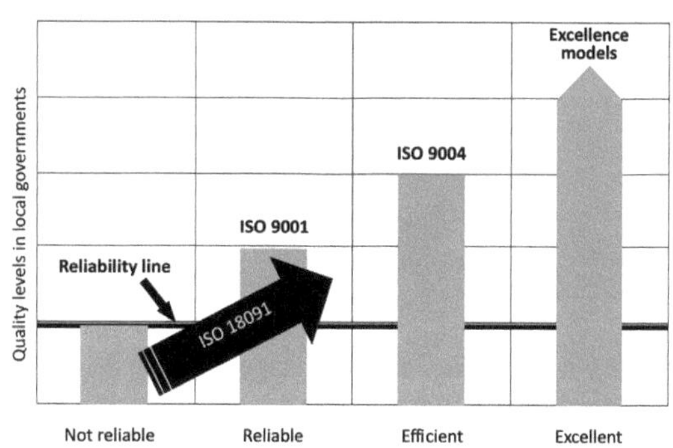

Fig. 1. Role of the standard *(*ISO 2022*)*ISO 18091:2022

ISO 18091 contains four annexes. Annex A, which is used in this research, provides a diagnostic model that can serve as a foundation for implementing a comprehensive quality management system within responsible local governance. Additionally, it includes Annexes B, C, and D.

For the research, its been used questionnaire found in Annex A of the ISO 18091. According to the ISO 18091 (2022) standard, this is a diagnostic system that can be used for self-evaluation and in this way the highest local government bodies have the possibility to identify the level of performance for each of indicators. This questionnaire contains 39 management indicators that are supported by about 130 United Nations documents and are classified into four groups of questions:

1. Institutional development for good governance;
2. Sustainable economic development;
3. Inclusive social development;
4. Sustainable development of thc environment.

These indicators include products/services that a typical local self-government should provide to its users - citizens in order to be considered reliable.

As shown in Fig. 2, there are three possible answers for each indicator: "red", "yellow" or "green". "Red" indicates essential practices are missing or are performed in an inadequate way by the local self-government to be considered reliable. "Yellow" cells indicate that the local government has made some efforts to implement the essential elements and is capable of providing the product/service as required. "Green" cells indicate that the minimum acceptable conditions have been met and that the local government can be considered reliable. Additionally, a "grey" level denotes best practice but isn't shown in the questionnaire, as this standard doesn't cover it. Fig. 3 illustrates the indicators.

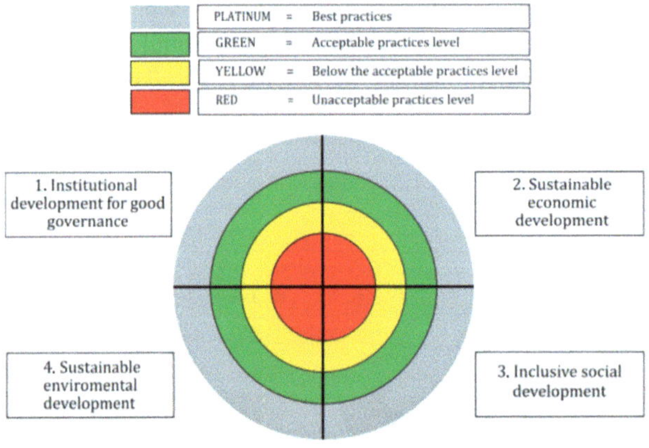

Fig. 2. Local government assessment tool for integral quality management (ISO 2022)

For the purposes of this research, it was taken a sample of 11 respondents from different municipalities in the Republic of Serbia. According to the data of the Ministry of State Administration and Local Self-Government, there are 174 units of local self-government in the Republic of Serbia. Thus, this sample constitutes 6% of the total population. The answers given by the respondents are shown on the graphs, while the indicators on the graphs are grouped according to groups of questions.

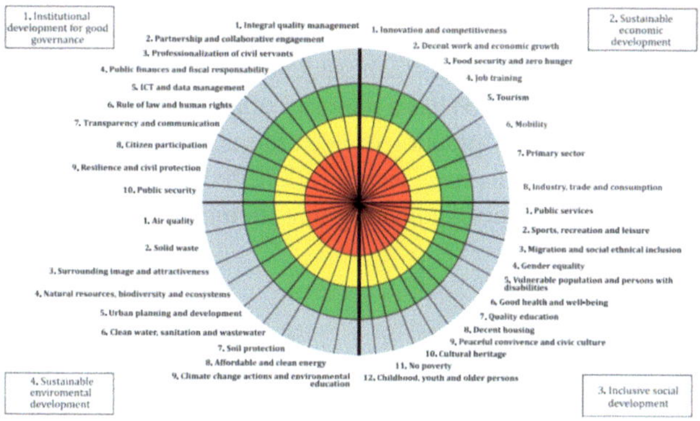

Fig. 3. Local government assessment tool for integrated management system (using 39 indicators) (ISO 2022)

5.1 RESULTS

First Group of Questions: Institutional Development for Good Governance

Figure 4 summarizes respondents' answers to the first group of questions. The "integral management system" indicator predominantly falls in the "yellow" category, reflecting practices below an acceptable level. In contrast, "partnership and collaborative arrangement" is largely in the "green" category, signifying acceptable performance. Responses for "professionalization of civil servants" are evenly split between "yellow" and "green." Indicators 4 through 8, addressing areas such as public finances and citizen participation, primarily fall in the "yellow" category, highlighting suboptimal performance. For "resilience and civil protection," responses are divided between "red" and "green," showing a mix of acceptable and unacceptable practices. Finally, Indicator 10 has the highest proportion of "red" responses, indicating an unacceptable level of practice across most municipalities.

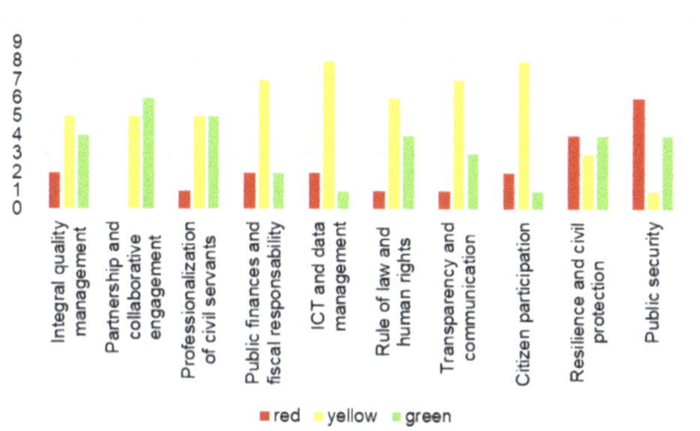

Fig. 4. Institutional development for good governance

The Second Group of Questions: Sustainable Economic Development

Figure 5 presents respondents' answers to the second group of questions. Among the indicators, only "tourism" (Indicator 5) has a majority of responses in the "green" category, reflecting an acceptable level of practice in most municipalities. In contrast, Indicators 1–4 and 6–8, which include areas such as innovation, decent work, and industry, primarily fall into the "yellow" category, indicating that municipalities generally fall short of acceptable practice levels.

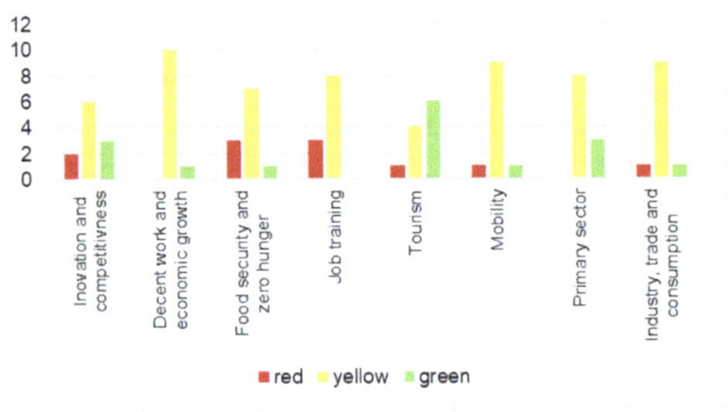

Fig. 5. Sustainable economic development

The Third Group of Questions: Inclusive Social Development

Figure 6 summarizes respondents' answers to the third group of questions. The first indicator, "public services," shows an even split between "yellow" and "green" responses, reflecting varying levels of practice among municipalities. The second indicator, "sports, recreation, and leisure," predominantly falls in the "green" category, indicating acceptable practice levels in most municipalities. Responses for "migration and socio-ethnic inclusion" are evenly divided between "red" and "green," highlighting significant disparities. Indicators such as "gender equality" and "vulnerable populations and persons with

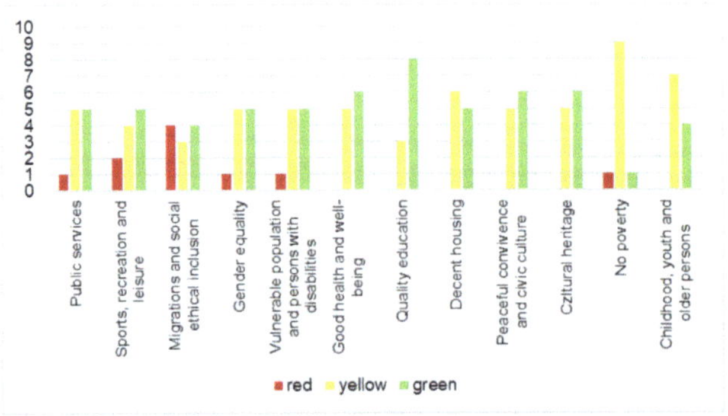

Fig. 6. Inclusive social development

disabilities" also show an equal split between "yellow" and "green." Most responses for indicators 6, 7, 9, and 10, including "good health and well-being" and "cultural heritage," fall in the "green" category, signaling acceptable practices. However, indicators 8, 11, and 12, covering areas like "dignified housing" and "childhood, youth, and the elderly," primarily fall into the "yellow" category, indicating areas where performance remains below acceptable standards.

The Fourth Group of Questions: Sustainable Development of The Environment
Figure 7 illustrates respondents' answers to the fourth group of questions. The first six indicators, including "air quality" and "urban planning," are predominantly in the "yellow" category, reflecting that most municipalities fall short of acceptable practice levels. "Soil protection" has the highest proportion of responses in the "red" category, indicating a significant level of unacceptable practice. Similarly, "affordable and clean energy" also shows a majority of "red" responses. The ninth indicator, "measures related to climate change and environmental education," mainly falls into the "yellow" category, highlighting below-standard performance in most municipalities (Fig. 8).

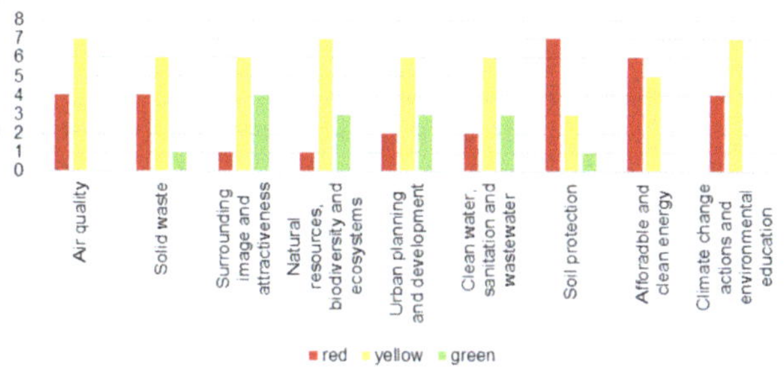

Fig. 7. Sustainable development of the environment

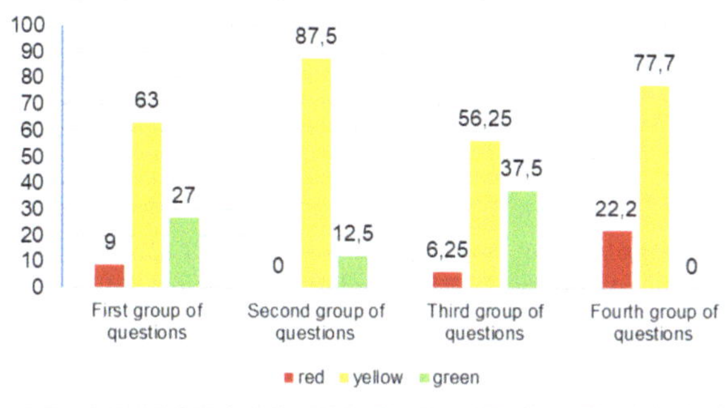

Fig. 8. Results summary in percentage

5.2 Discussion

Based on the results of the research, which included about 6% of organizations from the total population, it is possible to give certain guidelines for undertaking adequate activities aimed at fulfilling the requirements from Annex A of the ISO 18091 standard.

Institutional development for good governance, its better practice, in about two-thirds of organizations can achieve through a higher level of activities for integration of management systems, the professionalization of civil servants, public finance and fiscal responsibility, ICT and data management, the rule of law and human rights, transparency and communication and participation of citizens. Finally, a number of organizations, through the indicator of resilience and civil protection and public safety, show either a lack of conditions for their implementation or perform them in an inappropriate manner, which could be determined by additional research.

Sustainable economic development, a group of issues that opens up the opportunities to invest additional effort in all indicators and determine action guidelines, because it is obvious that organizations show a certain ability to maintain the processes in the mentioned indicators, but also that they are insufficient for the minimum acceptability of good practice.

Inclusive social development as a group of issues is the area that is obviously being worked on the most because organizations show the ability to carry out activities, but also that it is necessary to systematically raise the level through various new guidelines in order to reach the result of a minimum of good practices. Given that Inclusive Social Development actually means the inclusion of users, as well as participation, then the level of all services and activities should be minimally satisfied.

Sustainable development of the environment group of indicators, the results of which generally indicate in part the ability of local governments to undertake certain activities regarding all indicators, but that they are not at a sufficient level to qualify as a minimum of good practice. It is also indicative that soil protection and affordable and clean energy are in the zone of no activity or no adequate activity. Sustainable development of the environment refers either to an underdeveloped awareness of the environmental problems of the planet, our country and the local area, or to the absence of material and financial resources.

6 Conclusion

According to Thijs et al. (2017), public sector administrations are currently grappling with a triple challenge: delivering better services with fewer resources to meet social and business needs in times of tighter budgets; adapting service delivery to demographic, technological, and social changes; and improving the business environment through more efficient regulations and better services that foster growth and competitiveness.

In such a dynamic environment, the question of the quality of services that public sector institutions provide to citizens naturally arises.

As noted by the European Commission (2017a), "with approximately 75 million employees, the public sector is the largest single "industry" in Europe, employing around 25% of the workforce (with about 16% in central government alone) and accounting for nearly 50% of GDP. Given its scale, public administration — the organization and management of publicly funded resources — plays a critical role in citizens' daily lives and greatly impacts both performance and future prospects.".

One way to enhance sustainability in the public sector is to adopt integrated management systems (IMS), which can make organizations more flexible, effective, and efficient. Several IMS elements contribute to management efficiency, with the primary benefits being cost, time, and resource savings. IMS can be applied to any organization utilizing quality standards. Muzaimi et al. (2018) state that 'this system helps improve time management within an organization, as IMS implementation can reduce the time spent managing different systems, working hours, evaluation processes, and the time required for reviewing procedures and documentation.' Additionally, IMS enhances human resource management, and a well-implemented management system can improve the efficiency of everyday operations. The PDCA (Plan-Do-Check-Act) approach ensures the effective implementation of IMS and helps organizations improve their quality management practices.

In line with IMS implementation and its potential benefits, the European Commission (2017a:8) highlights that "it is essential for institutions managing economic and social interactions within the country to meet key criteria, such as the absence of corruption, a functional approach to competition and procurement policies, an effective legal framework, and a well-functioning judicial system. Additionally, strengthening institutional and administrative capacities, reducing administrative burdens, and improving legislative quality are vital for supporting structural adjustments, promoting economic growth, and fostering employment."

The public administration in the Republic of Serbia is currently undergoing reforms.

Research indicates that most local public administrations have created conditions for carrying out activities and providing services, but the effort needed to characterize these administrations as reliable remains insufficient. Although the research was conducted on a small sample of respondents, the findings clearly indicate the need for more comprehensive research to track reforms and involve a larger number of participants.

Therefore, public administration, as a large mechanism in every country, continuously interacts with citizens and faces challenges from a rapidly changing environment. According to Ruso and Filipovic (2020) "a strong driver for change in the Serbian legislation comes from its political decision to apply for EU membership. Many of the requirements laid down in international standards, directives and other legislative must be met by the Serbian economy so it can join the market mechanisms".

It must also find new ways to improve public sector performance and service delivery through organizational changes. This is why it is important that reforms are fully implemented — not only in technical, economic, political, and administrative terms but also through proper training and efforts from human resources, who are at the forefront of these challenges.

References

Almeida, D., Pradhan, N., Muniz Jr., J.: Assessment of ISO 9001:2015 implementation factors based on AHP: Case study in Brazilian automotive sector. Int. J. Qual. Reliab. Manag. **35**(7), 1343–1359 (2018). https://doi.org/10.1108/IJQRM-12-2016-0228

Almeida, J., Domingues, P., Sampaio, P.: Different perspectives on management systems integration. Total Qual. Manag. Bus. Excell. **25**(3–4), 338–351 (2014)
Andrews, R., Ferry, L., Skelcher, C., Wegorowski, P.: Corporatization in the public sector: Explaining the growth of local government companies. Public Adm. Rev. **80**(3), 482–493 (2020)
Anttila, J., Jussila, K.: ISO 9001: 2015–a questionable reform. What should the implementing organisations understand and do? Total Qual. Manag. Bus. Excell. **28**(9–10), 1090–1105 (2017)
Babeľová, Z.G., Stareček, A.: Evaluation of industrial enterprises' performance by different generations of employees. Entrep. Sustain. Issues. **9**(2), 346 (2021)
Barbosa, A.D.S., Bueno da Silva, L., de Souza, V.F., Morioka, S.N.: Integrated management systems: Their organizational impacts. Total Qual. Manag. Bus. Excell. **33**(7–8), 794–817 (2022)
Bencsik, A., Juhasz, T., Mura, L., Csanadi, A.: Impact of informal knowledge sharing for organizational operation. Entrep. Bus. Econ. Rev. **7**(3), 25–42 (2019)
Cho, J., Pyun, J.: A study on the implementation plan for public service quality management applying the ISO 18091 framework. J. Korean Soc. Qual. Manag. **50**(1), 1–19 (2022)
Council of Europe: European Charter of Local Self-Government. The Council of Europe, Strasbourg, France (1985) ETS 122 –
Ćwiklicki, M., Pawełek, B., Pilch, K.: Organisational resource capacity and ISO 9001 QMS Implementation in the Local Government. Evid. Pol., Public Organ. Rev. **21**(2), 205–219 (2021)
Dahlin, G., Isaksson, R.: Integrated management systems - interpretations, results, opportunities. TQM J. **29**(3), 528–542 (2017). https://doi.org/10.1108/TQM-01-2016-0004
Dick-Sagoe, C.: Decentralization for improving the provision of public services in developing countries: A critical review. Cogent Econ. Finance. **8**(1), 1804036 (2020)
European commission: Quality of Public Administration Toolbox: Principles and Values of Good Governance. Publications Office of the European Union, Luxembourg (2017a)
European commission: Quality of Public Administration Toolbox: Theme 3: Government Structures - Organization, Cooperation and Coordination. Publications Office of the European Union, Luxembourg (2017b)
European Commission's Inter-service group on Public Administration Quality and Innovation: Quality of Public Administration – A Toolbox for Practitioners. Publications Office of the European Union, Luxembourg (2017)
Gębczyńska, A.L.I.C.J.A., Wolniak, R.A.D.O.S.Ł.A.W.: Process Management Level in Local Government. CreativeSpace, Philadelphia (2018)
Goel, R.K., Mazhar, U., Nelson, M.A., Ram, R.: Different forms of decentralization and their impact on government performance: Micro-level evidence from 113 countries. Econ. Model. **62**, 171–183 (2017)
Ikram, M., Sroufe, R., Zhang, Q.: Prioritizing and overcoming barriers to integrated management system (IMS) implementation using AHP and G-TOPSIS. J. Clean. Prod. **254**, 120121 (2020)
Ingrams, A., Piotrowski, S., Berliner, D.: Learning from our mistakes: Public management reform and the hope of open government. Perspect. Public Manag. Gov. **3**(4), 257–272 (2020)
ISO: ISO 9001 in Local Government. International Organization for Standardization. [https://www.ISO.org/standard/45481.html, date of access: 01. 04. 2024.]ISO 18091:2022 - Quality management systems — Guidelines for the application of ISO 9001 in local government, Zurich, Switzerland (2022)
ISO: ISO Survey. Zurich, Switzerland: International Organization for Standardization. https://www.iso.org/committee/54998.html?t=KomURwikWDLiuB1P1c7SjLMLEAgXOA7emZHKGWyn8f3KQUTU3m287NxnpA3DIuxm&view=documents#section-isodocuments-top (2022a). Accessed 06 Apr 2024

Ispas, L., Mironeasa, C.: The identification of common models applied for the integration of management systems: A review. Sustainability. **14**(3559), 1–16 (2022). https://doi.org/10.3390/su14063559
ISS: Интерна правила стандардизације — Део 1: Доношење, објављивање, одржавање, преиспитивање и повлачење српских стандарда и сродних докумената. Institut za standardizaciju Srbije, Belgrade, Serbia (2023)
Karapetrovic, S., Willborn, W.: Integration of quality and environmental management systems. TQM Mag. **10**(3), 204–213 (1998)
Lapuente, V., Van de Walle, S.: The effects of new public management on the quality of public services. Governance. **33**(3), 461–475 (2020)
Lee, Y.W., Kim, J.M., Choi, S.Y., Park, K.B.: Obstacles in public sector TQM and the strategy to succeed: Application on Korea rehabilitation agency. J. Bus. Educ. **28**(6), 93–129 (2014)
Mavrot, C., Rosser, C., Sager, F., Hurni, P.Y.: A transfer-of-ideas approach to the history of public administration: The hybridizations of administrative traditions. In: Hildreth, B.W., Miller, J.G., Lindquist, A.E. (eds.) Handbook of Public Administration, pp. 13–24. Routledge, New York and London. (2021)
McAdam, R., Walker, T.: An inquiry into balanced scorecards within best value implementation in UK local government. Public Adm. **81**(4), 873–892 (2003)
Milošević, I., Ruso, J., Glogovac, M., Arsić, S., Rakić, A.: An integrated SEM-ANN approach for predicting QMS achievements in industry 4.0. Total Qual. Manag. Bus. Excell. **33**(15–16), 1896–1912 (2022)
Mizuno, K.: Leading by example: Local government in Japan adopts ISO 14000 and ISO 9000, funds SME implementation. ISO Manag. Syst. **2**(3), 21–28 (2002)
Muzaimi, H., Hamid, S.R., Chew, B.C.: Integrated management system for quality management system accreditation. J. Adv. Manuf. Technol. (JAMT). **12**(1 (1)), 87–100 (2018)
OECD: The Principles of Public Administration. OECD, Paris. https://www.sigmaweb.org/publications/Principles-of-Public-Administration-2023.pdf (2023). Accessed 21 May 2024
Olaru, M., Maier, D., Nicoară, D., Maier, A.: Establishing the basis for development of an organization by adopting the integrated management systems: Comparative study of various models and concepts of integration. Procedia Soc. Behav. Sci. **109**, 693–697 (2014)
European Parliament. (2024). Definition of principle of subsidiarity. The EU at work/ The European Union's legal system and decision-making procedures/The principle of subsidiarity
Pauliková, A., Chovancová, J., Blahová, J.: Cluster modeling of environmental and occupational health and safety management systems for integration support. Int. J. Environ. Res. Public Health. **19**(11), 6588 (2022)
Pinto, C., Domingues, J.P.T., Sampaio, P., Oliveira, O.: Integrated management systems in industry 4.0: Literature review. In: International Conference on Quality Engineering and Management (ICQEM), pp. 921–934. Riqual Publications, Braga, Portugal (2022)
Pittaway, J.J., Montazemi, A.R.: Know-how to lead digital transformation: The case of local governments. Gov. Inf. Q. **37**(4) p.1., 101474 (2020)
Rebelo, G., Santos, R.S.: Conception of a flexible integrator and lean model for integrated management systems. Total Qual. Manag. Bus. Excell. **25**(5–6), 683–701 (2014)
Ribeiro, F., Santos, G., Rebelo, M.F., Silva, R.: Integrated management systems: Trends for Portugal in the 2025 horizon. Procedia Manuf. **13**, 1191–1198 (2017)
Ruso, J., Djuric, M., Dimitrijević, S., Milošević, I.: Students' perception of quality management level in university student associations: The case of Serbia. Int. J. Qual. Res. **18**(4), 833–840 (2024)
Ruso, J., Filipovic, J.: How do public policy-makers perceive national quality infrastructure? The case of Serbia as an EU pre-accession country. Eur. Rev. **28**(2), 276–293 (2020)
Ruso, J., Krsmanovic, M., Trajkovic, A., Rakicevic, Z.: Quality management in public e-administration. Int. J. Econ. Manag. Eng. **7**(10), 2739–2743 (2013) pp. 550

Sfakianaki, E., Kakouris, A.P.: Obstacles to ISO 9001 certification in SMEs. Total Qual. Manag. Bus. Excell. **29**(1–2), 1–21 (2018). https://doi.org/10.1080/14783363.2018.1490640

Simon, A., Bernardo, M., Karapetrovic, S., Casadesus, M.: Implementing integrated management systems in chemical firms. Total Qual. Manag. Bus. Excell. **24**(3–4), 294–309 (2013)

Šućurović, A., Milošević, J.K., Terečik, E.: Investigating the implementation of integrated management system (IMS). Acta Tech. Corviniensis-Bull. Eng. **17**(2), 59–62 (2024)

Talapatra, S., Santos, G., Sharf Uddin, K., Carvalho, F.: Main benefits of integrated management systems through literature review. On Qual. Innov. Sustain. **13**(4), 85–97 (2019)

The British Standards Institution. PAS 99 Integrated Management System. http://www.bsigroup.com/en-AE/PAS-99- Integrated-Management-Systems/ (2016). Accessed 12 Aug 2024

Thijs, N., Hammerschmid, G., Palaric, E.: A comparative overview of public administration characteristics and performance in EU28. Eur. Comm., Brussels (2017)

Tolo, M.L.: The effect of ISO 9001: 2015 on public service quality: the case of a Gauteng provincial government department. Doctoral dissertation. Cape Peninsula University of Technology (2023)

Trajković, A., Milošević, I.: Model to determine the economic and other effects of standardisation – a case study in Serbia. Total Qual. Manag. Bus. Excell. **29**(5–6), 673–685 (2016). https://doi.org/10.1080/14783363.2016.1225496

Treaty of Lisbon: Amending the treaty on European union and the treaty establishing the European community, protocol on the application of the principles of subsidiarity and proportionality. Off. J. Eur. Union. **50**, 150–153. https://eur-lex.europa.eu/legal-content/EN/TXT/PDF/?uri=OJ:C:2007:306:FULL- (2007).. Accessed 28 Apr 2024

Vukosavljević, Đ., Kaputo, J., Tešić, A., Vukosavljević, D.: Makroekonomsko okruženje i javni sektor–mesto i uloga. Oditor. **7, br. 3**, 37–50 (2021)

Zakon. Zakon o lokalnoj samoupravi. “Službeni glasnik RS", br. 129/2007, 83/2014 - dr. zakon, 101/2016 - dr. zakon, 47/2018 i 111/2021 - dr. zakon (2021)

Author Index

V. Štavljanin et al. (Eds.): SymOrg 2024, LNNS 1681, p. 213, 2025.
https://doi.org/10.1007/978-3-032-08090-5

MIX
Papier aus verantwortungsvollen Quellen
Paper from responsible sources
FSC® C105338

If you have any concerns about our products,
you can contact us on
ProductSafety@springernature.com

In case Publisher is established outside the EU,
the EU authorized representative is:
Springer Nature Customer Service Center GmbH
Europaplatz 3, 69115 Heidelberg, Germany

Printed by Libri Plureos GmbH
in Hamburg, Germany